ALSO BY STEPHEN JAY GOULD

Ever Since Darwin: Reflections in Natural History

Ontogeny and Phylogeny

The Panda's Thumb: More Reflections in Natural History

A View of Life (with S. E. Luria and S. Singer)

THE MISMEASURE OF MAN

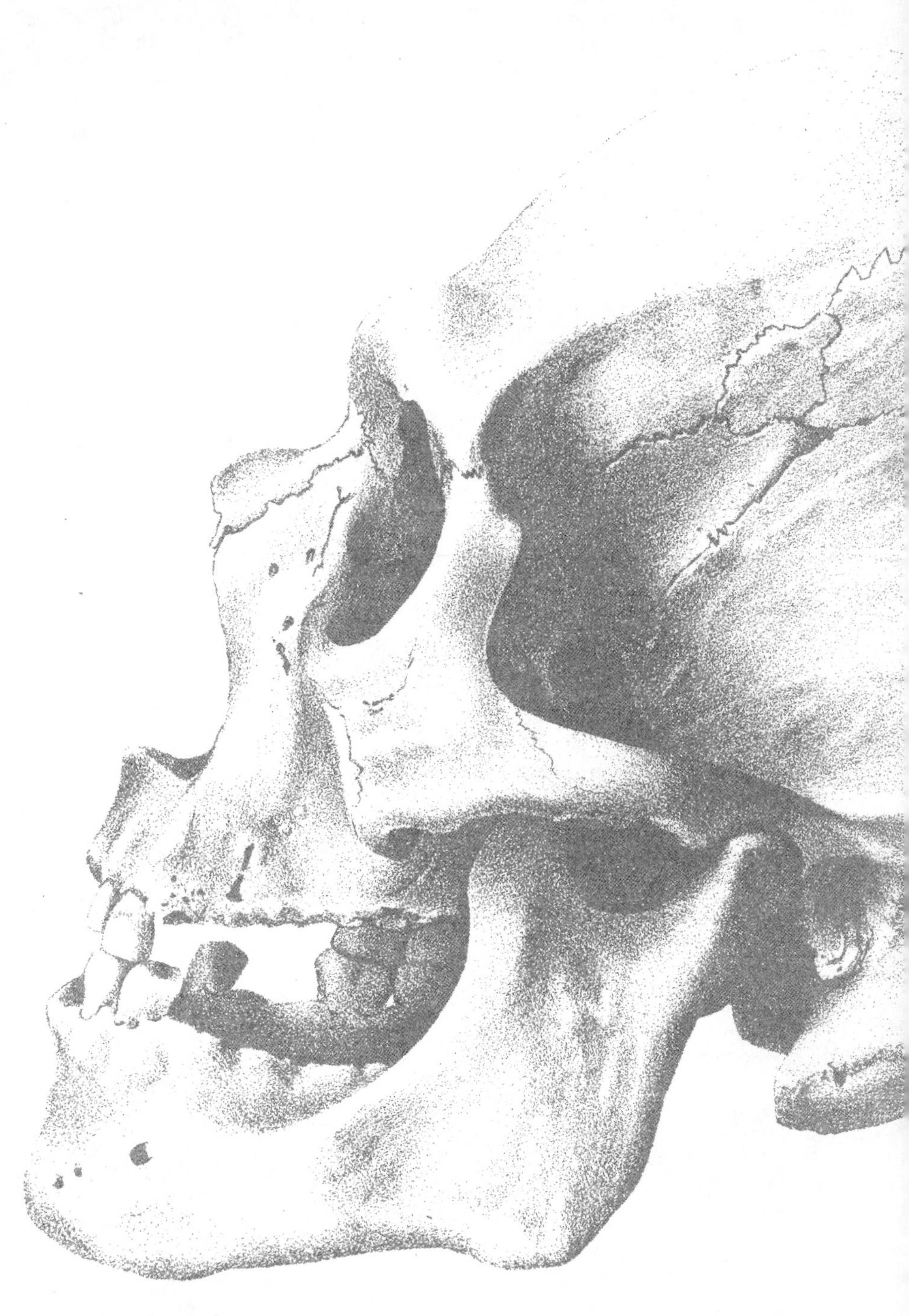

THE Mismeasure of Man

BY STEPHEN JAY GOULD

If the misery of our poor be caused not by the laws of nature, but by our institutions, great is our sin. —Charles Darwin, *Voyage of the Beagle*

W · W · NORTON & COMPANY · *NEW YORK* · *LONDON*

THE TEXT OF THIS BOOK is composed in photocomposition Baskerville. Display type is Typositor Caslon 540.

BOOK DESIGN BY MARJORIE J. FLOCK

Library of Congress Cataloging in Publication Data
Gould, Stephen Jay.
The mismeasure of man.
Bibliography: p. 337
Includes index.
1. Intelligence tests—History. 2. Ability—Testing—History. 3. Personality tests—History. 4. Craniometry—History. I. Title.
BF431.G68 1981 153.9'3 81-38430 AACR2

ISBN 0-393-01489-4

W. W. Norton & Company, Inc. 500 Fifth Avenue, New York, N.Y. 10110
W. W. Norton & Company Ltd. 37 Great Russell Street, London, W.C. 1

To the memory of Grammy and Papa Joe,
who came, struggled, and prospered,
Mr. Goddard notwithstanding.

Contents

6. The Real Error of Cyril Burt: *Factor Analysis and the Reification of Intelligence* 234

Acknowledgments

GENES MAY BE SELFISH in a limited metaphorical sense, but there can be no gene for selfishness when I have so many friends and colleagues willing to offer their aid. I thank Ashley Montagu, not only for his specific suggestions, but also for leading the fight against scientific racism for so many years without becoming cynical about human possibilities. Several colleagues who have written, or are writing, their own books on biological determinism willingly shared their information and even let me use their own findings, sometimes before they could publish them themselves: G. Allen, A. Chase, S. Chorover, L. Kamin, R. Lewontin. Others heard of my efforts and, without solicitation, sent material and suggestions that enriched the book greatly: M. Leitenberg, S. Selden. L. Meszoly prepared the original illustrations in Chapter 6. Perhaps Kropotkin was right after all; I shall remain with the hopeful.

A note on references: In place of conventional footnotes, I have used the system of references universally found in scientific literature—name of author and year of publication, cited in parentheses after the relevant passage of text. (Items are then listed by author and by year for any one author in the bibliography.) I know that many readers may be disconcerted at first; the text will seem cluttered to many. Yet, I am confident that everyone will begin to "read through" the citations after a few pages of experience, and will then discover that they do not interrupt the flow of prose. To me, the advantages of this system far outweigh any aesthetic deficit—no more flipping back and forth from text to end-notes (no publisher will set them all at the bottom of the page any more), only to find that a tantalizing little number yields no juicy tidbit of subsidiary information, but only a dry bibliographic citation;* immediate access to the two essential bits of information for any historical inquiry—who and when. I believe that this system of referencing is one of the few potential contributions that scientists, normally not a very literate lot, might supply to other fields of written scholarship.

A note on title: I hope that an apparently sexist title will be taken in the intended spirit—not only as a play on Protagoras' famous aphorism, but also as a commentary on the procedures of biological determinists discussed in the book. They did, indeed, study "man" (that is, white European males), regarding this group as a standard and everybody else as something to be measured unfavorably against it. That they mismeasured "man" underscores the double fallacy.

*The relatively small number of truly informational footnotes can then be placed at the bottom of the page, where they belong.

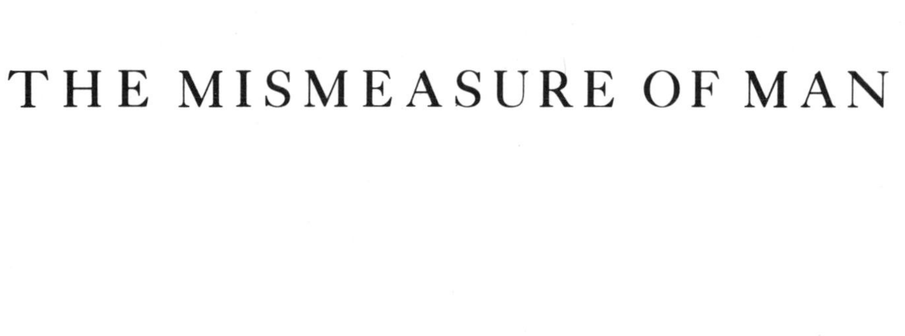

THE MISMEASURE OF MAN

ONE

Introduction

CITIZENS OF THE REPUBLIC, Socrates advised, should be educated and assigned by merit to three classes: rulers, auxiliaries, and craftsmen. A stable society demands that these ranks be honored and that citizens accept the status conferred upon them. But how can this acquiescence be secured? Socrates, unable to devise a logical argument, fabricates a myth. With some embarrassment, he tells Glaucon:

> I will speak, although I really know not how to look you in the face, or in what words to utter the audacious fiction. . . . They [the citizens] are to be told that their youth was a dream, and the education and training which they received from us, an appearance only; in reality during all that time they were being formed and fed in the womb of the earth. . . .

Glaucon, overwhelmed, exclaims: "You had good reason to be ashamed of the lie which you were going to tell." "True," replied Socrates, "but there is more coming; I have only told you half."

> Citizens, we shall say to them in our tale, you are brothers, yet God has framed you differently. Some of you have the power of command, and in the composition of these he has mingled gold, wherefore also they have the greatest honor; others he has made of silver, to be auxiliaries; others again who are to be husbandmen and craftsmen he has composed of brass and iron; and the species will generally be preserved in the children. . . . An oracle says that when a man of brass or iron guards the State, it will be destroyed. Such is the tale; is there any possibility of making our citizens believe in it?

Glaucon replies: "Not in the present generation; there is no way of accomplishing this; but their sons may be made to believe in the tale, and their son's sons, and posterity after them."

Glaucon had uttered a prophesy. The same tale, in different versions, has been promulgated and believed ever since. The justification for ranking groups by inborn worth has varied with the tides of Western history. Plato relied upon dialectic, the Church upon dogma. For the past two centuries, scientific claims have become the primary agent for validating Plato's myth.

This book is about the scientific version of Plato's tale. The general argument may be called *biological determinism*. It holds that shared behavioral norms, and the social and economic differences between human groups—primarily races, classes, and sexes—arise from inherited, inborn distinctions and that society, in this sense, is an accurate reflection of biology. This book discusses, in historical perspective, a principal theme within biological determinism: the claim that worth can be assigned to individuals and groups by *measuring intelligence as a single quantity*. Two major sources of data have supported this theme: craniometry (or measurement of the skull) and certain styles of psychological testing.

Metals have ceded to genes (though we retain an etymological vestige of Plato's tale in speaking of people's worthiness as their "mettle"). But the basic argument has not changed: that social and economic roles accurately reflect the innate construction of people. One aspect of the intellectual strategy has altered, however. Socrates knew that he was telling a lie.

Determinists have often invoked the traditional prestige of science as objective knowledge, free from social and political taint. They portray themselves as purveyors of harsh truth and their opponents as sentimentalists, ideologues, and wishful thinkers. Louis Agassiz (1850, p. 111), defending his assignment of blacks to a separate species, wrote: "Naturalists have a right to consider the questions growing out of men's physical relations as merely scientific questions, and to investigate them without reference to either politics or religion." Carl C. Brigham (1923), arguing for the exclusion of southern and eastern European immigrants who had scored poorly on supposed tests of innate intelligence stated: "The steps that should be taken to preserve or increase our present intellectual capacity must of course be dictated by science and not by political expediency." And Cyril Burt, invoking faked data compiled by the nonexistent Ms. Conway, complained that doubts about the genetic foundation of IQ "appear to be based rather on the social ideals or the subjective preferences of the critics than on any first-hand examination of the evidence supporting the opposite view" (in Conway, 1959, p. 15).

Since biological determinism possesses such evident utility for groups in power, one might be excused for suspecting that it also arises in a political context, despite the denials quoted above. After all, if the status quo is an extension of nature, then any major change, if possible at all, must inflict an enormous cost—psychological for individuals, or economic for society—in forcing people into unnatural arrangements. In his epochal book, *An American Dilemma* (1944), Swedish sociologist Gunnar Myrdal discussed the thrust of biological and medical arguments about human nature: "They have been associated in America, as in the rest of the world, with conservative and even reactionary ideologies. Under their long hegemony, there has been a tendency to assume biological causation without question, and to accept social explanations only under the duress of a siege of irresistible evidence. In political questions, this tendency favored a do-nothing policy." Or, as Condorcet said more succinctly a long time ago: they "make nature herself an accomplice in the crime of political inequality."

This book seeks to demonstrate both the scientific weaknesses and political contexts of determinist arguments. Even so, I do not intend to contrast evil determinists who stray from the path of scientific objectivity with enlightened antideterminists who approach data with an open mind and therefore see truth. Rather, I criticize the myth that science itself is an objective enterprise, done properly only when scientists can shuck the constraints of their culture and view the world as it really is.

Among scientists, few conscious ideologues have entered these debates on either side. Scientists needn't become explicit apologists for their class or culture in order to reflect these pervasive aspects of life. My message is not that biological determinists were bad scientists or even that they were always wrong. Rather, I believe that science must be understood as a social phenomenon, a gutsy, human enterprise, not the work of robots programed to collect pure information. I also present this view as an upbeat for science, not as a gloomy epitaph for a noble hope sacrificed on the altar of human limitations.

Science, since people must do it, is a socially embedded activity.

It progresses by hunch, vision, and intuition. Much of its change through time does not record a closer approach to absolute truth, but the alteration of cultural contexts that influence it so strongly. Facts are not pure and unsullied bits of information; culture also influences what we see and how we see it. Theories, moreover, are not inexorable inductions from facts. The most creative theories are often imaginative visions imposed upon facts; the source of imagination is also strongly cultural.

This argument, although still anathema to many practicing scientists, would, I think, be accepted by nearly every historian of science. In advancing it, however, I do not ally myself with an overextension now popular in some historical circles: the purely relativistic claim that scientific change only reflects the modification of social contexts, that truth is a meaningless notion outside cultural assumptions, and that science can therefore provide no enduring answers. As a practicing scientist, I share the credo of my colleagues: I believe that a factual reality exists and that science, though often in an obtuse and erratic manner, can learn about it. Galileo was not shown the instruments of torture in an abstract debate about lunar motion. He had threatened the Church's conventional argument for social and doctrinal stability: the static world order with planets circling about a central earth, priests subordinate to the Pope and serfs to their lord. But the Church soon made its peace with Galileo's cosmology. They had no choice; the earth really does revolve about the sun.

Yet the history of many scientific subjects is virtually free from such constraints of fact for two major reasons. First, some topics are invested with enormous social importance but blessed with very little reliable information. When the ratio of data to social impact is so low, a history of scientific attitudes may be little more than an oblique record of social change. The history of scientific views on race, for example, serves as a mirror of social movements (Provine, 1973). This mirror reflects in good times and bad, in periods of belief in equality and in eras of rampant racism. The death knell of the old eugenics in America was sounded more by Hitler's particular use of once-favored arguments for sterilization and racial purification than by advances in genetic knowledge.

Second, many questions are formulated by scientists in such a restricted way that any legitimate answer can only validate a social preference. Much of the debate on racial differences in mental worth, for example, proceeded upon the assumption that intelligence is a thing in the head. Until this notion was swept aside, no amount of data could dislodge a strong Western tradition for ordering related items into a progressive chain of being.

Science cannot escape its curious dialectic. Embedded in surrounding culture, it can, nonetheless, be a powerful agent for questioning and even overturning the assumptions that nurture it. Science can provide information to reduce the ratio of data to social importance. Scientists can struggle to identify the cultural assumptions of their trade and to ask how answers might be formulated under different assertions. Scientists can propose creative theories that force startled colleagues to confront unquestioned procedures. But science's potential as an instrument for identifying the cultural constraints upon it cannot be fully realized until scientists give up the twin myths of objectivity and inexorable march toward truth. One must, indeed, locate the beam in one's own eye before interpreting correctly the pervasive motes in everybody else's. The beams can then become facilitators, rather than impediments.

Gunnar Myrdal (1944) captured both sides of this dialectic when he wrote:

> A handful of social and biological scientists over the last 50 years have gradually forced informed people to give up some of the more blatant of our biological errors. But there must be still other countless errors of the same sort that no living man can yet detect, because of the fog within which our type of Western culture envelops us. Cultural influences have set up the assumptions about the mind, the body, and the universe with which we begin; pose the questions we ask; influence the facts we seek; determine the interpretation we give these facts; and direct our reaction to these interpretations and conclusions.

Biological determinism is too large a subject for one man and one book—for it touches virtually every aspect of the interaction between biology and society since the dawn of modern science. I have therefore confined myself to one central and manageable argument in the edifice of biological determinism—an argument in two historical chapters, based on two deep fallacies, and carried forth in one common style.

The argument begins with one of the fallacies—*reification,* or our tendency to convert abstract concepts into entities (from the Latin *res,* or thing). We recognize the importance of mentality in our lives and wish to characterize it, in part so that we can make the divisions and distinctions among people that our cultural and political systems dictate. We therefore give the word "intelligence" to this wondrously complex and multifaceted set of human capabilities. This shorthand symbol is then reified and intelligence achieves its dubious status as a unitary thing.

Once intelligence becomes an entity, standard procedures of science virtually dictate that a location and physical substrate be sought for it. Since the brain is the seat of mentality, intelligence must reside there.

We now encounter the second fallacy—*ranking,* or our propensity for ordering complex variation as a gradual ascending scale. Metaphors of progress and gradualism have been among the most pervasive in Western thought—see Lovejoy's classic essay (1936) on the great chain of being or Bury's famous treatment (1920) of the idea of progress. Their social utility should be evident in the following advice from Booker T. Washington (1904, p. 245) to black America:

> For my race, one of its dangers is that it may grow impatient and feel that it can get upon its feet by artificial and superficial efforts rather than by the slower but surer process which means one step at a time through all the constructive grades of industrial, mental, moral and social development which all races have had to follow that have become independent and strong.

But ranking requires a criterion for assigning all individuals to their proper status in the single series. And what better criterion than an objective number? Thus, the common style embodying both fallacies of thought has been quantification, or the measurement of intelligence as a single number for each person.* This book, then, is about the abstraction of intelligence as a single entity, its location within the brain, its quantification as one number for

*Peter Medawar (1977, p. 13) has presented other interesting examples of "the illusion embodied in the ambition to attach a single number valuation to complex quantities"—for example, the attempts made by demographers to seek causes for trends in population in a single measure of "reproductive prowess," or the desire of soil scientists to abstract the "quality" of a soil as a single number.

each individual, and the use of these numbers to rank people in a single series of worthiness, invariably to find that oppressed and disadvantaged groups—races, classes, or sexes—are innately inferior and deserve their status. In short, this book is about the Mismeasure of Man.*

Different arguments for ranking have characterized the last two centuries. Craniometry was the leading numerical science of biological determinism during the nineteenth century. I discuss (Chapter 2) the most extensive data compiled before Darwin to rank races by the sizes of their brains—the skull collection of Philadelphia physician Samuel George Morton. Chapter 3 treats the flowering of craniometry as a rigorous and respectable science in the school of Paul Broca in late nineteenth-century Europe. Chapter 4 then underscores the impact of quantified approaches to human anatomy in nineteenth-century biological determinism. It presents two case studies: the theory of recapitulation as evolution's primary criterion for unilinear ranking of human groups, and the attempt to explain criminal behavior as a biological atavism reflected in the apish morphology of murderers and other miscreants.

What craniometry was for the nineteenth century, intelligence testing has become for the twentieth, when it assumes that intelligence (or at least a dominant part of it) is a single, innate, heritable, and measurable thing. I discuss the two components of this invalid approach to mental testing in Chapter 5 (the hereditarian version of the IQ scale as an American product) and Chapter 6 (the argument for reifying intelligence as a single entity by the mathematical technique of factor analysis). Factor analysis is a difficult mathematical subject almost invariably omitted from documents written for nonprofessionals. Yet I believe that it can be made accessible and explained in a pictorial and nonnumerical way. The material of Chapter 6 is still not "easy reading," but I could not leave it out—for the history of intelligence testing cannot be understood without grasping the factor analytic argument and understanding its deep

*Following strictures of the argument outlined above, I do not treat all theories of craniometrics (I omit phrenology, for example, because it did not reify intelligence as a single entity but sought multiple organs with the brain). Likewise, I exclude many important and often quantified styles of determinism that did not seek to measure intelligence as a property of the brain—for example, most of eugenics.

conceptual fallacy. The great IQ debate makes no sense without this conventionally missing subject.

I have tried to treat these subjects in an unconventional way by using a method that falls outside the traditional purview of either a scientist or historian operating alone. Historians rarely treat the quantitative details in sets of primary data. They write, as I cannot adequately, about social context, biography, or general intellectual history. Scientists are used to analyzing the data of their peers, but few are sufficiently interested in history to apply the method to their predecessors. Thus, many scholars have written about Broca's impact, but no one has recalculated his sums.

I have focused upon the reanalysis of classical data sets in craniometry and intelligence testing for two reasons beyond my incompetence to proceed in any other fruitful way and my desire to do something a bit different. I believe, first of all, that Satan also dwells with God in the details. If the cultural influences upon science can be detected in the humdrum minutiae of a supposedly objective, almost automatic quantification, then the status of biological determinism as a social prejudice reflected by scientists in their own particular medium seems secure.

The second reason for analyzing quantitative data arises from the special status that numbers enjoy. The mystique of science proclaims that numbers are the ultimate test of objectivity. Surely we can weigh a brain or score an intelligence test without recording our social preferences. If ranks are displayed in hard numbers obtained by rigorous and standardized procedures, then they must reflect reality, even if they confirm what we wanted to believe from the start. Antideterminists have understood the particular prestige of numbers and the special difficulty that their refutation entails. Léonce Manouvrier (1903, p. 406), the nondeterminist black sheep of Broca's fold, and a fine statistician himself, wrote of Broca's data on the small brains of women:

> Women displayed their talents and their diplomas. They also invoked philosophical authorities. But they were opposed by *numbers* unknown to Condorcet or to John Stuart Mill. These numbers fell upon poor women like a sledge hammer, and they were accompanied by commentaries and sarcasms more ferocious than the most misogynist imprecations of certain church fathers. The theologians had asked if women had a soul. Several centuries later, some scientists were ready to refuse them a human intelligence.

If—as I believe I have shown—quantitative data are as subject to cultural constraint as any other aspect of science, then they have no special claim upon final truth.

In reanalyzing these classical data sets, I have continually located a priori prejudice, leading scientists to invalid conclusions from adequate data, or distorting the gathering of data itself. In a few cases—Cyril Burt's documented fabrication of data on IQ of identical twins, and my discovery that Goddard altered photographs to suggest mental retardation in the Kallikaks—we can specify conscious fraud as the cause of inserted social prejudice. But fraud is not historically interesting except as gossip because the perpetrators know what they are doing and the *unconscious* biases that record subtle and inescapable constraints of culture are not illustrated. In most cases discussed in this book, we can be fairly certain that biases—though often expressed as egregiously as in cases of conscious fraud—were unknowingly influential and that scientists believed they were pursuing unsullied truth.

Since many of the cases presented here are so patent, even risible, by today's standards, I wish to emphasize that I have not taken cheap shots at marginal figures (with the possible exceptions of Mr. Bean in Chapter 3, whom I use as a curtain-raiser to illustrate a general point, and Mr. Cartwright in Chapter 2, whose statements are too precious to exclude). Cheap shots come in thick catalogues—from a eugenicist named W. D. McKim, Ph.D. (1900), who thought that all nocturnal housebreakers should be dispatched with carbonic acid gas, to a certain English professor who toured the United States during the late nineteenth century, offering the unsolicited advice that we might solve our racial problems if every Irishman killed a Negro and got hanged for it.* Cheap shots are also gossip, not history; they are ephemeral and uninfluential, however amusing. I have focused upon the leading and most influential scientists of their times and have analyzed their major works.

I have enjoyed playing detective in most of the case studies that make up this book: finding passages expurgated without comment

*Also too precious to exclude is my favorite modern invocation of biological determinism as an excuse for dubious behavior. Bill Lee, baseball's self-styled philosopher, justifying the beanball (*New York Times,* 24 July 1976): "I read a book in college called 'Territorial Imperative.' A fellow always has to protect his master's home much stronger than anything down the street; My territory is down and away from the hitters. If they're going out there and getting the ball, I'll have to come in close."

in published letters, recalculating sums to locate errors that support expectations, discovering how adequate data can be filtered through prejudices to predetermined results, even giving the Army Mental Test for illiterates to my own students with interesting results. But I trust that whatever zeal any investigator must invest in details has not obscured the general message: that determinist arguments for ranking people according to a single scale of intelligence, no matter now numerically sophisticated, have recorded little more than social prejudice—and that we learn something hopeful about the nature of science in pursuing such an analysis.

If this subject were merely a scholar's abstract concern, I could approach it in more measured tone. But few biological subjects have had a more direct influence upon millions of lives. Biological determinism is, in its essence, a *theory of limits.* It takes the current status of groups as a measure of where they should and must be (even while it allows some rare individuals to rise as a consequence of their fortunate biology).

I have said little about the current resurgence of biological determinism because its individual claims are usually so ephemeral that their refutation belongs in a magazine article or newspaper story. Who even remembers the hot topics of ten years ago: Shockley's proposals for reimbursing voluntarily sterilized individuals according to their number of IQ points below 100, the great XYY debate, or the attempt to explain urban riots by diseased neurology of rioters. I thought that it would be more valuable and interesting to examine the original sources of the arguments that still surround us. These, at least, display great and enlightening errors. But I was inspired to write this book because biological determinism is rising in popularity again, as it always does in times of political retrenchment. The cocktail party circuit has been buzzing with its usual profundity about innate aggression, sex roles, and the naked ape. Millions of people are now suspecting that their social prejudices are scientific facts after all. Yet these latent prejudices themselves, not fresh data, are the primary source of renewed attention.

We pass through this world but once. Few tragedies can be more extensive than the stunting of life, few injustices deeper than the denial of an opportunity to strive or even to hope, by a limit imposed from without, but falsely identified as lying within. Cicero tells the story of Zopyrus, who claimed that Socrates had inborn vices evident in his physiognomy. His disciples rejected the claim, but Socrates defended Zopyrus and stated that he did indeed possess the vices, but had cancelled their effects through the exercise of reason. We inhabit a world of human differences and predilections, but the extrapolation of these facts to theories of rigid limits is ideology.

George Eliot well appreciated the special tragedy that biological labeling imposed upon members of disadvantaged groups. She expressed it for people like herself—women of extraordinary talent. I would apply it more widely—not only to those whose dreams are flouted but also to those who never realize that they may dream. But I cannot match her prose (from the prelude to *Middlemarch*):

> Some have felt that these blundering lives are due to the inconvenient indefiniteness with which the Supreme Power has fashioned the natures of women: if there were one level of feminine incompetence as strict as the ability to count three and no more, the social lot of women might be treated with scientific certitude. The limits of variation are really much wider than anyone would imagine from the sameness of women's coiffure and the favorite love stories in prose and verse. Here and there a cygnet is reared uneasily among the ducklings in the brown pond, and never finds the living stream in fellowship with its own oary-footed kind. Here and there is born a Saint Theresa, foundress of nothing, whose loving heartbeats and sobs after an unattained goodness tremble off and are dispersed among hindrances instead of centering in some long-recognizable deed.

TWO

American Polygeny and Craniometry before Darwin

Blacks and Indians as Separate, Inferior Species

Order is Heaven's first law; and, this confessed,
Some are, and must be, greater than the rest.
—ALEXANDER POPE, *Essay on Man* (1733)

APPEALS TO REASON or to the nature of the universe have been used throughout history to enshrine existing hierarchies as proper and inevitable. The hierarchies rarely endure for more than a few generations, but the arguments, refurbished for the next round of social institutions, cycle endlessly.

The catalogue of justifications based on nature traverses a range of possibilities: elaborate analogies between rulers and a hierarchy of subordinate classes with the central earth of Ptolemaic astronomy and a ranked order of heavenly bodies circling around it; or appeals to the universal order of a "great chain of being," ranging in a single series from amoebae to God, and including near its apex a graded series of human races and classes. To quote Alexander Pope again:

Without this just gradation, could they be
Subjected, these to those, or all to thee?
....
From Nature's chain whatever link you strike,
Tenth, or ten thousandth, breaks the chain alike.

The humblest, as well as the greatest, play their part in preserving the continuity of universal order; all occupy their appointed roles.

This book treats an argument that, to many people's surprise, seems to be a latecomer: biological determinism, the notion that people at the bottom are constructed of intrinsically inferior material (poor brains, bad genes, or whatever). Plato, as we have seen, cautiously floated this proposal in the *Republic*, but finally branded it as a lie.

Racial prejudice may be as old as recorded human history, but its biological justification imposed the additional burden of intrinsic inferiority upon despised groups, and precluded redemption by conversion or assimilation. The "scientific" argument has formed a primary line of attack for more than a century. In discussing the first biological theory supported by extensive quantitative data—early nineteenth-century craniometry—I must begin by posing a question of causality: did the introduction of inductive science add legitimate data to change or strengthen a nascent argument for racial ranking? Or did a priori commitment to ranking fashion the "scientific" questions asked and even the data gathered to support a foreordained conclusion?

A shared context of culture

In assessing the impact of science upon eighteenth- and nineteenth-century views of race, we must first recognize the cultural milieu of a society whose leaders and intellectuals did not doubt the propriety of racial ranking—with Indians below whites, and blacks below everybody else (Fig. 2.1). Under this universal umbrella, arguments did not contrast equality with inequality. One group—we might call them "hard-liners"—held that blacks were inferior and that their biological status justified enslavement and colonization. Another group—the "soft-liners," if you will—agreed that blacks were inferior, but held that a people's right to freedom did not depend upon their level of intelligence. "Whatever be their degree of talents," wrote Thomas Jefferson, "it is no measure of their rights."

Soft-liners held various attitudes about the nature of black disadvantage. Some argued that proper education and standard of life could "raise" blacks to a white level; others advocated permanent black ineptitude. They also disagreed about the biological or cultural roots of black inferiority. Yet, throughout the egalitarian tradition of the European Enlightenment and the American revolution, I cannot identify any popular position remotely like the "cultural relativism" that prevails (at least by lip-service) in liberal circles today. The nearest approach is a common argument that black inferiority is purely cultural and that it can be completely eradicated by education to a Caucasian standard.

All American culture heroes embraced racial attitudes that would embarrass public-school mythmakers. Benjamin Franklin, while viewing the inferiority of blacks as purely cultural and completely remediable, nonetheless expressed his hope that America would become a domain of whites, undiluted by less pleasing colors.

> I could wish their numbers were increased. And while we are, as I may call it, scouring our planet, by clearing America of woods, and so making this side of our globe reflect a brighter light to the eyes of inhabitants in Mars or Venus, why should we . . . darken its people? Why increase the Sons of Africa, by planting them in America, where we have so fair an opportunity, by excluding all blacks and tawneys, of increasing the lovely white and red?* (*Observations Concerning the Increase of Mankind,* 1751).

Others among our heroes argued for biological inferiority. Thomas Jefferson wrote, albeit tentatively: "I advance it, therefore, as a suspicion only, that the blacks, whether originally a distinct race, or made distinct by time and circumstance, are inferior to the whites in the endowment both of body and of mind" (in Gossett,

* I have been struck by the frequency of such aesthetic claims as a basis of racial preference. Although J. F. Blumenbach, the founder of anthropology, had stated that toads must view other toads as paragons of beauty, many astute intellectuals never doubted the equation of whiteness with perfection. Franklin at least had the decency to include the original inhabitants in his future America; but, a century later, Oliver Wendell Holmes rejoiced in the elimination of Indians on aesthetic grounds: ". . . and so the red-crayon sketch is rubbed out, and the canvas is ready for a picture of manhood a little more like God's own image" (in Gossett, 1965, p. 243).

---

2 • 1 The unilinear scale of human races and lower relatives according to Nott and Gliddon, 1868. The chimpanzee skull is falsely inflated, and the Negro jaw extended, to give the impression that blacks might even rank lower than the apes.

Apollo Belvidere

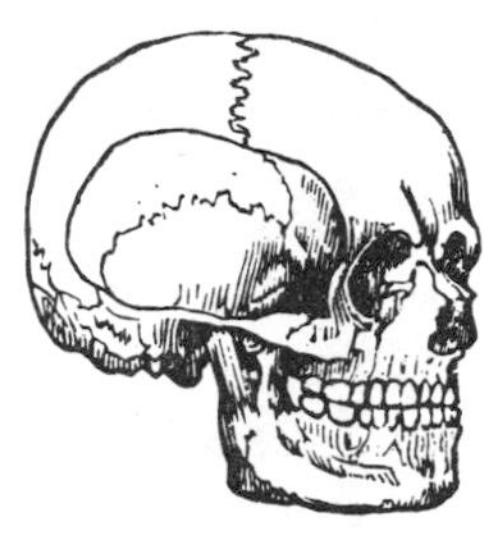

Greek

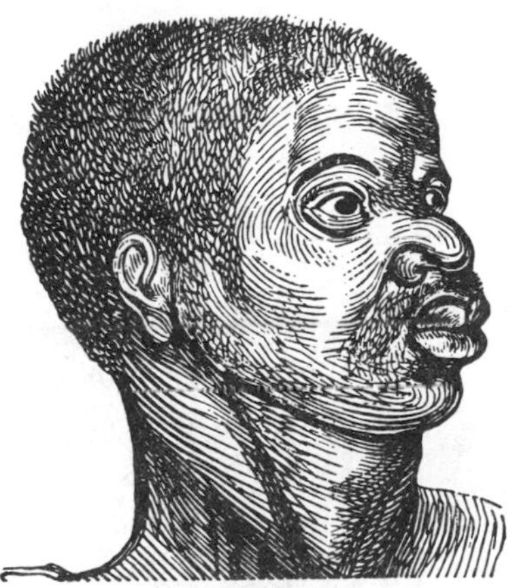

Negro

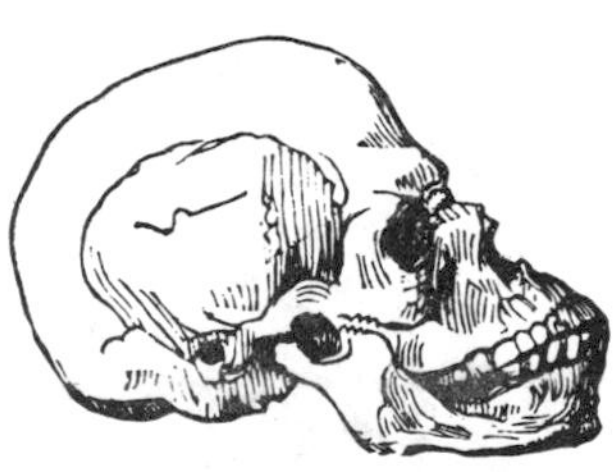

Creole Negro

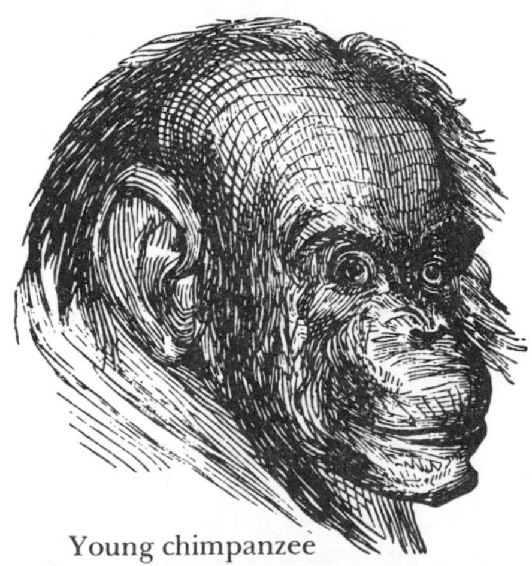

Young chimpanzee

Young chimpanzee

Algerian Negro　　　　Saharran Negro

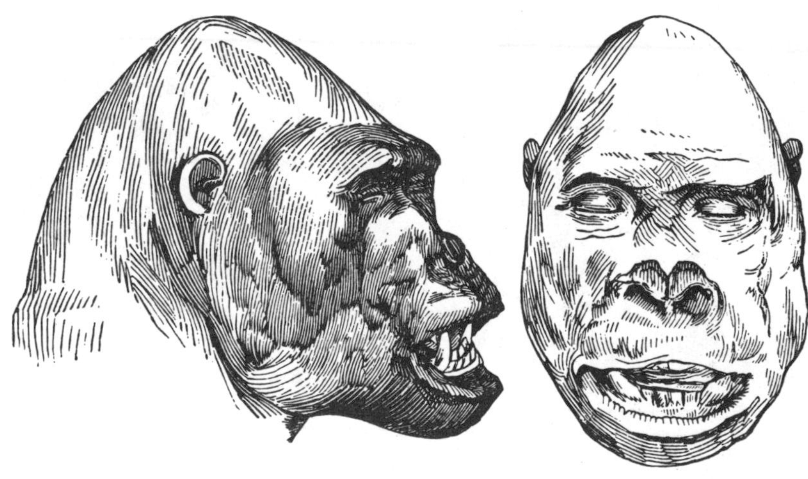

Gorilla

2 • 2 An unsubtle attempt to suggest strong affinity between blacks and gorillas. From Nott and Gliddon, *Types of Mankind,* 1854. Nott and Gliddon comment on this figure: "The palpable analogies and dissimilitudes between an inferior type of mankind and a superior type of monkey require no comment."

1965, p. 44). Lincoln's pleasure at the performance of black soldiers in the Union army greatly increased his respect for freedmen and former slaves. But freedom does not imply biological equality, and Lincoln never abandoned a basic attitude, so strongly expressed in the Douglas debates (1858):

> There is a physical difference between the white and black races which I believe will forever forbid the two races living together on terms of social and political equality. And inasmuch as they cannot so live, while they do remain together there must be the position of superior and inferior, and I as much as any other man am in favor of having the superior position assigned to the white race.

Lest we choose to regard this statement as mere campaign rhetoric, I cite this private jotting, scribbled on a fragment of paper in 1859:

> Negro equality! Fudge! How long, in the Government of a God great enough to make and rule the universe, shall there continue knaves to vend, and fools to quip, so low a piece of demagogism as this (in Sinkler, 1972, p. 47).

I do not cite these statements in order to release skeletons from ancient closets. Rather, I quote the men who have justly earned our highest respect in order to show that white leaders of Western nations did not question the propriety of racial ranking during the eighteenth and nineteenth centuries. In this context, the pervasive assent given by scientists to conventional rankings arose from shared social belief, not from objective data gathered to test an open question. Yet, in a curious case of reversed causality, these pronouncements were read as independent support for the political context.

All leading scientists followed social conventions (Figs. 2.2 and 2.3). In the first formal definition of human races in modern taxonomic terms, Linnaeus mixed character with anatomy (*Systema naturae,* 1758). *Homo sapiens afer* (the African black), he proclaimed, is "ruled by caprice"; *Homo sapiens europaeus* is "ruled by customs." Of African women, he wrote: *Feminis sine pudoris; mammae lactantes prolixae*—Women without shame, breasts lactate profusely. The men, he added, are indolent and annoint themselves with grease.

The three greatest naturalists of the nineteenth century did not hold blacks in high esteem. Georges Cuvier, widely hailed in France as the Aristotle of his age, and a founder of geology, paleontology, and modern comparative anatomy, referred to native Africans as "the most degraded of human races, whose form approaches that of the beast and whose intelligence is nowhere great enough to arrive at regular government" (Cuvier, 1812, p. 105). Charles Lyell, the conventional founder of modern geology, wrote:

The brain of the Bushman . . . leads towards the brain of the Simiadae [monkeys]. This implies a connexion between want of intelligence and structural assimilation. Each race of Man has its place, like the inferior animals (in Wilson, 1970, p. 347).

Charles Darwin, the kindly liberal and passionate abolitionist,* wrote about a future time when the gap between human and ape will increase by the anticipated extinction of such intermediates as chimpanzees and Hottentots.

The break will then be rendered wider, for it will intervene between man in a more civilized state, as we may hope, than the Caucasian, and some ape as low as a baboon, instead of as at present between the negro or Australian and the gorilla (*Descent of Man,* 1871, p. 201).

Even more instructive are the beliefs of those few scientists often cited in retrospect as cultural relativisits and defenders of equality. J. F. Blumenbach attributed racial differences to the influences of climate. He protested rankings based on beauty or presumed mental ability and assembled a collection of books written by blacks. Nonetheless, he did not doubt that white people set

* Darwin wrote, for example, in the *Voyage of the Beagle:* "Near Rio de Janeiro I lived opposite to an old lady, who kept screws to crush the fingers of her female slaves. I have stayed in a house where a young household mulatto, daily and hourly, was reviled, beaten, and persecuted enough to break the spirit of the lowest animal. I have seen a little boy, six or seven years old, struck thrice with a horse-whip (before I could interfere) on his naked head, for having handed me a glass of water not quite clean. . . . And these deeds are done and palliated by men, who profess to love their neighbors as themselves, who believe in God, and pray that his Will be done on earth! It makes one's blood boil, yet heart tremble, to think that we Englishmen and our American descendants, with their boastful cry of liberty, have been and are so guilty."

2•3 Two more comparisons of blacks and apes from Nott and Gliddon, 1854. This book was not a fringe document, but the leading American text on human racial differences.

Orangutan

Hottentot wagoner

Chimpanzee

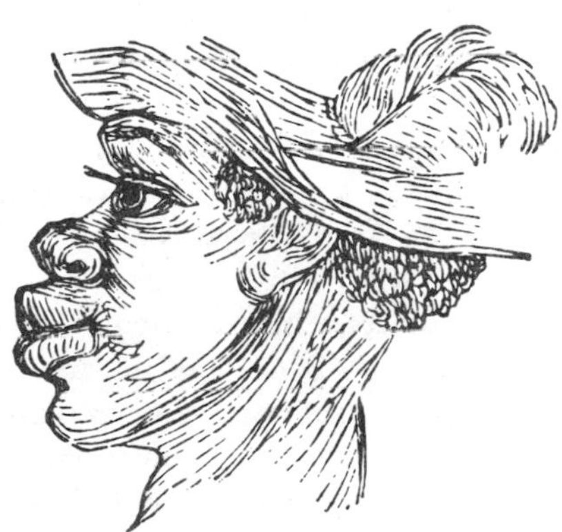

Hottentot from Somerset

a standard, from which all other races must be viewed as departures:

> The Caucasian must, on every physiological principle, be considered as the primary or intermediate of these five principal Races. The two extremes into which it has deviated, are on the one hand the Mongolian, on the other the Ethiopian [African blacks] (1825, p. 37).

Alexander von Humboldt, world traveler, statesman, and greatest popularizer of nineteenth-century science, would be the hero of all modern egalitarians who seek antecedents in history. He, more than any other scientist of his time, argued forcefully and at length against ranking on mental or aesthetic grounds. He also drew political implications from his convictions, and campaigned against all forms of slavery and subjugation as impediments to the natural striving of all people to attain mental excellence. He wrote in the most famous passage of his five-volume *Cosmos:*

> Whilst we maintain the unity of the human species, we at the same time repel the depressing assumption of superior and inferior races of men. There are nations more susceptible of cultivation than others—but none in themselves nobler than others. All are in like degree designed for freedom (1849, p. 368).

Yet even Humboldt invoked innate mental difference to resolve some dilemmas of human history. Why, he asks in the second volume of *Cosmos,* did the Arabs explode in culture and science soon after the rise of Islam, while Scythian tribes of southeastern Europe stuck to their ancient ways; for both peoples were nomadic and shared a common climate and environment? Humboldt did find some cultural differences—greater contact of Arabs with surrounding urbanized cultures, for example. But, in the end, he labeled Arabs as a "more highly gifted race" with greater "natural adaptability for mental cultivation" (1849, p. 578).

Alfred Russel Wallace, codiscoverer of natural selection with Darwin, is justly hailed as an antiracist. Indeed, he did affirm near equality in the innate mental capacity of all peoples. Yet, curiously, this very belief led him to abandon natural selection and return to divine creation as an explanation for the human mind—much to Darwin's disgust. Natural selection, Wallace argued, can only build structures immediately useful to animals possessing them. The brain of savages is, potentially, as good as ours. But they do not use it fully, as the rudeness and inferiority of their culture indicates. Since modern savages are much like human ancestors, our brain must have developed its higher capacities long before we put them to any use.

Preevolutionary styles of scientific racism: monogenism and polygenism

Preevolutionary justifications for racial ranking proceeded in two modes. The "softer" argument—again using some inappropriate definitions from modern perspectives—upheld the scriptural unity of all peoples in the single creation of Adam and Eve. This view was called *monogenism*—or origin from a single source. Human races are a product of degeneration from Eden's perfection. Races have declined to different degrees, whites least and blacks most. Climate proved most popular as a primary cause for racial distinction. Degenerationists differed on the remediability of modern deficits. Some held that the differences, though developed gradually under the influence of climate, were now fixed and could never be reversed. Others argued that the fact of gradual development implied reversibility in appropriate environments. Samuel Stanhope Smith, president of the College of New Jersey (later Princeton), hoped that American blacks, in a climate more suited to Caucasian temperaments, would soon turn white. But other degenerationists felt that improvement in benevolent climes could not proceed rapidly enough to have any impact upon human history.

The "harder" argument abandoned scripture as allegorical and held that human races were separate biological species, the descendants of different Adams. As another form of life, blacks need not participate in the "equality of man." Proponents of this argument were called "polygenists."

Degenerationism was probably the more popular argument, if only because scripture was not to be discarded lightly. Moreoever, the interfertility of all human races seemed to guarantee their union as a single species under Buffon's criterion that members of a species be able to breed with each other, but not with representatives of any other group. Buffon himself, the greatest of eighteenth-century France, was a strong abolitionist and exponent of improvement for inferior races in appropriate environments. But he never doubted the inherent validity of a white standard:

> The most temperate climate lies between the 40th and 50th degree of latitude, and it produces the most handsome and beautiful men. It is from this climate that the ideas of the genuine color of mankind, and of the various degrees of beauty ought to be derived.

Some degenerationists cited their commitments in the name of human brotherhood. Etienne Serres, a famous French medical anatomist, wrote in 1860 that the perfectability of lower races distinguished humans as the only species subject to improvement by its own efforts. He lambasted polygeny as a "savage theory" that "seems to lend scientific support to the enslavement of races less advanced in civilization than the Caucasian":

> Their conclusion is that the Negro is no more a white man than a donkey is a horse or a zebra—a theory put into practice in the United States of America, to the shame of civilization (1860, pp. 407–408).

Nonetheless, Serres worked to document the signs of inferiority among lower races. As an anatomist, he sought evidence within his specialty and confessed to some difficulty in establishing both criteria and data. He settled on the theory of recapitulation—the idea that higher creatures repeat the adult stages of lower animals during their own growth (Chapter 4). Adult blacks, he argued, should be like white children, adult Mongolians like white adolescents. He searched diligently but devised nothing much better than the distance between navel and penis—"that ineffaceable sign of embryonic life in man." This distance is small relative to body height in babies of all races. The navel migrates upward during growth, but attains greater heights in whites than in yellows, and never gets very far at all in blacks. Blacks remain perpetually like white children and announce their inferiority thereby.

Polygeny, though less popular, had its illustrious supporters as well. David Hume did not spend his life absorbed in pure thought. He held a number of political posts, including the stewardship of the English colonial office in 1766. Hume advocated both the separate creation and innate inferiority of nonwhite races:

> I am apt to suspect the negroes and in general all the other species of men (for there are four or five different kinds) to be naturally inferior to

the whites. There never was a civilized nation of any other complexion than white, nor even any individual eminent either in action or speculation.* No ingenious manufacturers amongst them, no arts, no sciences. . . . Such a uniform and constant difference could not happen in so many countries and ages, if nature had not made an original distinction betwixt these breeds of men. Not to mention our colonies, there are negroes slaves dispersed all over Europe, of which none ever discovered any symptoms of ingenuity, tho' low people without education will start up amongst us, and distinguish themselves in every profession. In Jamaica indeed they talk of one negroe as a man of parts and learning; but 'tis likely he is admired for very slender accomplishments like a parrot who speaks a few words plainly (in Popkin, 1974, p. 143; see Popkin's excellent article for a long analysis of Hume as a polygenist).

Charles White, an English surgeon, wrote the strongest defense of polygeny in 1799—*Account of the Regular Gradation in Man.* White abandoned Buffon's criterion of interfertility in defining species, pointing to successful hybrids between such conventionally separate groups as foxes, wolves, and jackals.† He railed against the idea that climate might produce racial differences, arguing that such ideas might lead, by extension, to the "degrading notion" of evolution between species. He disclaimed any political motivation and announced an untainted purpose: "to investigate a proposition in natural history." He explicitly rejected any extension of polygeny to "countenance the pernicious practice of enslaving man-

* This "inductive" argument from human cultures is far from dead as a defense of racism. In his *Study of History* (1934 edition), Arnold Toynbee wrote: "When we classify mankind by color, the only one of the primary races, given by this classification, which has not made a creative contribution to any of our twenty-one civilizations is the Black Race" (in Newby, 1969, p. 217).

† Modern evolutionary theory does invoke a barrier to interfertility as the primary criterion for status as a species. In the standard definition: "Species are actually or potentially interbreeding populations sharing a common gene pool, and reproductively isolated from all other groups." Reproductive isolation, however, does not mean that individual hybrids never arise, but only that the two species maintain their integrity in natural contact. Hybrids may be sterile (mules). Fertile hybrids may even arise quite frequently, but if natural selection acts preferentially against them (as a result of inferiority in structural design, rejection as mates by full members of either species, etc.) they will not increase in frequency and the two species will not amalgamate. Often fertile hybrids can be produced in the laboratory by imposing situations not encountered in nature (forced breeding between species that normally mature at different times of the year, for example). Such examples do not refute a status as separate species because the two groups do not amalgamate in the wild (maturation at different times of the year may be an efficient means of reproductive isolation).

kind." White's criteria of ranking tended toward the aesthetic, and his argument included the following gem, often quoted. Where else but among Caucasians, he argued, can we find

> . . . that nobly arched head, containing such a quantity of brain. . . . Where that variety of features, and fulness of expression; those long, flowing, graceful ring-lets; that majestic beard, those rosy cheeks and coral lips? Where that . . . noble gait? In what other quarter of the globe shall we find the blush that overspreads the soft features of the beautiful women of Europe, that emblem of modesty, of delicate feelings . . . where, except on the bosom of the European woman, two such plump and snowy white hemispheres, tipt with vermillion (in Stanton, 1960, p. 17).

Louis Agassiz—America's theorist of polygeny

Ralph Waldo Emerson argued that intellectual emancipation should follow political independence. American scholars should abandon their subservience to European styles and theories. We have, Emerson wrote, "listened too long to the courtly muses of Europe." "We will walk on our own feet; we will work with our own hands; we will speak our own minds" (in Stanton, 1960, p. 84).

In the early to mid-nineteenth century, the budding profession of American science organized itself to follow Emerson's advice. A collection of eclectic amateurs, bowing before the prestige of European theorists, became a group of professionals with indigenous ideas and an internal dynamic that did not require constant fueling from Europe. The doctrine of polygeny acted as an important agent in this transformation; for it was one of the first theories of largely American origin that won the attention and respect of European scientists—so much so that Europeans referred to polygeny as the "American school" of anthropology. Polygeny had European antecedents, as we have seen, but Americans developed the data cited in its support and based a large body of research on its tenets. I shall concentrate on the two most famous advocates of polygeny—Agassiz the theorist and Morton the data analyst; and I shall try to uncover both the hidden motives and the finagling of data so central to their support.* For starters, it is obviously not accidental that a nation still practicing slavery and expelling its aboriginal inhabitants from their homelands should have provided

* An excellent history of the entire "American school" can be found in W. Stanton's *The Leopard's Spots.*

a base for theories that blacks and Indians are separate species, inferior to whites.

Louis Agassiz (1807–1873), the great Swiss naturalist, won his reputation in Europe, primarily as Cuvier's disciple and a student of fossil fishes. His immigration to America in the 1840s immediately elevated the status of American natural history. For the first time, a major European theorist had found enough of value in the United States to come and stay. Agassiz became a professor at Harvard, where he founded and directed the Museum of Comparative Zoology until his death in 1873 (I occupy an office in the original wing of his building). Agassiz was a charmer; he was lionized in social and intellectual circles from Boston to Charlestown. He spoke for science with boundless enthusiasm and raised money with equal zeal to support his buildings, collections, and publications. No man did more to establish and enhance the prestige of American biology during the nineteenth century.

Agassiz also became the leading spokesman for polygeny in America. He did not bring this theory with him from Europe. He converted to the doctrine of human races as separate species after his first experiences with American blacks.

Agassiz did not embrace polygeny as a conscious political doctrine. He never doubted the propriety of racial ranking, but he did count himself among the opponents of slavery. His adherence to polygeny flowed easily from procedures of biological research that he had developed in other and earlier contexts. He was, first of all, a devout creationist who lived long enough to become the only major scientific opponent of evolution. But nearly all scientists were creationists before 1859, and most did not become polygenists (racial differentiation within a single species posed no threat to the doctrine of special creation—just consider breeds of dogs and cattle). Agassiz's predisposition to polygeny arose primarily from two aspects of his personal theories and methods:

1. In studying the geographic distribution of animals and plants, Agassiz developed a theory about "centers of creation." He believed that species were created in their proper places and did not generally migrate far from these centers. Other biogeographers invoked creation in a single spot with extensive migration thereafter. Thus, when Agassiz studied what we would now regard as a single widespread species, divided into fairly distinct geographical races, he tended to name several separate species, each

created at its center of origin. *Homo sapiens* is a primary example of a cosmopolitan, variable species.

2. Agassiz was an extreme splitter in his taxonomic practice. Taxonomists tend to fall into two camps—"lumpers," who concentrate on similarities and amalgamate groups with small differences into single species, and "splitters," who focus on minute distinctions and establish species on the smallest peculiarities of design. Agassiz was a splitter among splitters. He once named three genera of fossil fishes from isolated teeth that a later paleontologist found in the variable dentition of a single individual. He named invalid species of freshwater fishes by the hundreds, basing them upon peculiar individuals within single, variable species. An extreme splitter who viewed organisms as created over their entire range might well be tempted to regard human races as separate creations. Nonetheless, before coming to America, Agassiz advocated the doctrine of human unity—even though he viewed our variation as exceptional. He wrote in 1845:

> Here is revealed anew the superiority of the human genre and its greater independence in nature. Whereas the animals are distinct species in the different zoological provinces to which they appertain, man, despite the diversity of his races, constitutes one and the same species over all the surface of the globe (in Stanton, 1960, p. 101).

Agassiz may have been predisposed to polygeny by biological belief, but I doubt that this pious man would have abandoned the Biblical orthodoxy of a single Adam if he had not been confronted both by the sight of American blacks and the urgings of his polygenist colleagues. Agassiz never generated any data for polygeny. His conversion followed an immediate visceral judgment and some persistent persuasion by friends. His later support rested on nothing deeper in the realm of biological knowledge.

Agassiz had never seen a black person in Europe. When he first met blacks as servants at his Philadelphia hotel in 1846, he experienced a pronounced visceral revulsion. This jarring experience, coupled with his sexual fears about miscegenation, apparently established his conviction that blacks are a separate species. In a remarkably candid passage, he wrote to his mother from America:

> It was in Philadelphia that I first found myself in prolonged contact with negroes; all the domestics in my hotel were men of color. I can scarcely

express to you the painful impression that I received, especially since the feeling that they inspired in me is contrary to all our ideas about the confraternity of the human type [*genre*] and the unique origin of our species. But truth before all. Nevertheless, I experienced pity at the sight of this degraded and degenerate race, and their lot inspired compassion in me in thinking that they are really men. Nonetheless, it is impossible for me to repress the feeling that they are not of the same blood as us. In seeing their black faces with their thick lips and grimacing teeth, the wool on their head, their bent knees, their elongated hands, their large curved nails, and especially the livid color of the palm of their hands, I could not take my eyes off their face in order to tell them to stay far away. And when they advanced that hideous hand towards my plate in order to serve me, I wished I were able to depart in order to eat a piece of bread elsewhere, rather than dine with such service. What unhappiness for the white race—to have tied their existence so closely with that of negroes in certain countries! God preserve us from such a contact! (Agassiz to his mother, December 1846.) (The standard *Life and Letters,* compiled by Agassiz's wife, omits these lines in presenting an expurgated version of this famous letter. Other historians have paraphrased them or passed them by. I recovered this passage from the original manuscript in Harvard's Houghton Library and have translated it, verbatim, for the first time so far as I know.)

Agassiz published his major statement on human races in the *Christian Examiner* for 1850. He begins by dismissing as demagogues both the divines who would outlaw him as an infidel (for preaching the doctrine of multiple Adams) and the abolitionists who would brand him as a defender of slavery:

> It has been charged upon the views here advanced that they tend to the support of slavery. . . . Is that a fair objection to a philosophical investigation? Here we have to do only with the question of the origin of men; let the politicians, let those who feel themselves called upon to regulate human society, see what they can do with the results. . . . We disclaim, however, all connection with any question involving political matters. It is simply with reference to the possibility of appreciating the differences existing between different men, and of eventually determining whether they have originated all over the world, and under what circumstances, that we have here tried to trace some facts respecting the human races (1850, p. 113).

Agassiz then presents his argument: The theory of polygeny does not constitute an attack upon the scriptural doctrine of human unity. Men are bound by a common structure and sympathy, even though races were created as separate species. The Bible does not speak about parts of the world unknown to the ancients; the tale of Adam refers only to the origin of Caucasians. Negroes and Caucasians are as distinct in the mummified remains of Egypt as they are today. If human races were the product of climatic influence, then the passage of three thousand years would have engendered substantial changes (Agassiz had no inkling of human antiquity; he believed that three thousand years included a major chunk of our entire history). Modern races occupy definite, non-overlapping, geographic areas—even though some ranges have been blurred or obliterated by migration. As physically distinct, temporally invariant groups with discrete geographical ranges, human races met all Agassiz's biological criteria for separate species.

> These races must have originated . . . in the same numerical proportions, and over the same area, in which they now occur. . . . They cannot have originated in single individuals, but must have been created in that numeric harmony which is characteristic of each species; men must have originated in nations, as the bees have originated in swarms (pp. 128–129).

Then, approaching the end of his article, Agassiz abruptly shifts his ground and announces a moral imperative—even though he had explicitly justified his inquiry by casting it as an objective investigation of natural history.

> There are upon earth different races of men, inhabiting different parts of its surface, which have different physical characters; and this fact . . . presses upon us the obligation to settle the relative rank among these races, the relative value of the characters peculiar to each, in a scientific point of view. . . . As philosophers it is our duty to look it in the face (p. 142).

As direct evidence for differential, innate value Agassiz ventures no further than the standard set of Caucasian cultural stereotypes:

> The indominable, courageous, proud Indian—in how very different a light he stands by the side of the submissive, obsequious, imitative negro, or by the side of the tricky, cunning, and cowardly Mongolian! Are not these facts indications that the different races do not rank upon one level in nature (p. 144).

Blacks, Agassiz declares, must occupy the bottom rung of any objective ladder:

> It seems to us to be mock-philanthropy and mock-philosophy to assume that all races have the same abilities, enjoy the same powers, and show the same natural dispositions, and that in consequence of this equality they are entitled to the same position in human society. History speaks here for itself. . . . This compact continent of Africa exhibits a population which has been in constant intercourse with the white race, which has enjoyed the benefit of the example of the Egyptian civilization, of the Phoenician civilization, of the Roman civilization, of the Arab civilization . . . and nevertheless there has never been a regulated society of black men developed on that continent. Does not this indicate in this race a peculiar apathy, a peculiar indifference to the advantages afforded by civilized society? (pp. 143–144).

If Agassiz had not made his political message clear, he ends by advocating specific social policy. Education, he argues, must be tailored to innate ability; train blacks in hand work, whites in mind work:

> What would be the best education to be imparted to the different races in consequence of their primitive difference, . . . We entertain not the slightest doubt that human affairs with reference to the colored races would be far more judiciously conducted if, in our intercourse with them, we were guided by a full consciousness of the real difference existing between us and them, and a desire to foster those dispositions that are eminently marked in them, rather than by treating them on terms of equality (p. 145).

Since those "eminently marked" dispositions are submissiveness, obsequiousness, and imitation, we can well imagine what Agassiz had in mind. I have treated this paper in detail because it is so typical of its genre—advocacy of social policy couched as a dispassionate inquiry into scientific fact. The strategy is by no means moribund today.

In a later correspondence, pursued in the midst of the Civil War, Agassiz expressed his political views more forcefully and at greater length. (These letters are also expurgated without indication in the standard version published by Agassiz's wife. Again, I have restored passages from the original letters in Harvard's Houghton Library.) S. G. Howe, a member of Lincoln's Inquiry Commission, asked Agassiz's opinion about the role of blacks in a reunited nation. (Howe, known best for his work in prison reform and education of the blind, was the husband of Julia Ward Howe, author of the "Battle Hymn of the Republic".) In four long and impassioned letters, Agassiz pleaded his case. The persistence of a large and permanent black population in America must be acknowledged as a grim reality. Indians, driven by their commendable pride, may perish in battle, but "the negro exhibits by nature a pliability, a readiness to accommodate himself to circumstances, a proneness to imitate those among whom he lives" (9 August 1863).

Although legal equality must be granted to all, blacks should be denied social equality, lest the white race be compromised and diluted: "Social equality I deem at all time impracticable. It is a natural impossibility flowing from the very character of the negro race" (10 August 1863); for blacks are "indolent, playful, sensuous, imitative, subservient, good natured, versatile, unsteady in their purpose, devoted, affectionate, in everything unlike other races, they may but be compared to children, grown in the stature of adults while retaining a childlike mind. . . . Therefore I hold that they are incapable of living on a footing of social equality with the whites, in one and the same community, without being an element of social disorder" (10 August 1863). Blacks must be regulated and limited, lest an injudicious award of social privilege sow later discord:

> No man has a right to what he is unfit to use. . . . Let us beware of granting too much to the negro race in the beginning, lest it become necessary to recall violently some of the privileges which they may use to our detriment and their own injury (10 August 1863).

For Agassiz, nothing inspired more fear than the prospect of amalgamation by intermarriage. White strength depends upon separation: "The production of halfbreeds is as much a sin against nature, as incest in a civilized community is a sin against purity of character. . . . Far from presenting to me a natural solution of our difficulties, the idea of amalgamation is most repugnant to my feelings, I hold it to be a perversion of every natural sentiment. . . . No efforts should be spared to check that which is abhorrent to our better nature, and to the progress of a higher civilization and a purer morality" (9 August 1863).

Agassiz now realizes that he has argued himself into a corner. If interbreeding among races (separate species to Agassiz) is unnatural and repugnant, why are "halfbreeds" so common in America?

Agassiz attributes this lamentable fact to the sexual receptiveness of housemaids and the naïveté of young Southern gentlemen. The servants, it seems, are halfbreeds already (we are not told how their parents overcame a natural repugnance for one another); young men respond aesthetically to the white half, while a degree of black heritage loosens the natural inhibitions of a higher race. Once acclimated, the poor young men are hooked, and they acquire a taste for pure blacks:

> As soon as the sexual desires are awakening in the young men of the South, they find it easy to gratify themselves by the readiness with which they are met by colored [halfbreed] house servants. . . . This blunts his better instincts in that direction and leads him gradually to seek more spicy partners, as I have heard the full blacks called by fast young men (9 August 1863).

Finally, Agassiz combines vivid image and metaphor to warn against the ultimate danger of a mixed and enfeebled people:

> Conceive for a moment the difference it would make in future ages, for the prospect of republican institutions and our civilization generally, if instead of the manly population descended from cognate nations the United States should hereafter be inhabited by the effeminate progeny of mixed races, half indian, half negro, sprinkled with white blood. . . . I shudder from the consequences. We have already to struggle, in our progress, against the influence of universal equality, in consequence of the difficulty of preserving the acquisitions of individual eminence, the wealth of refinement and culture growing out of select associations. What would be our condition if to these difficulties were added the far more tenacious influences of physical disability. . . . How shall we eradicate the stigma of a lower race when its blood has once been allowed to flow freely into that of our children (10 August 1863).*

Agassiz concludes that legal freedom awarded to slaves in manumission must spur the enforcement of rigid social separation among races. Fortunately, nature shall be the accomplice of moral

*E. D. Cope, America's leading paleontologist and evolutionary biologist, reiterated the same theme even more forcefully in 1890 (p. 2054): "The highest race of man cannot afford to lose or even to compromise the advantages it has acquired by hundreds of centuries of toil and hardship, by mingling its blood with the lowest. . . . We cannot cloud or extinguish the fine nervous susceptibility, and the mental force, which cultivation develops in the constitution of the Indo-European, by the fleshly instincts, and dark mind of the African. Not only is the mind stagnated, and the life of mere living introduced in its stead, but the possibility of resurrection is rendered doubtful or impossible."

virtue; for people, free to choose, gravitate naturally toward the climates of their original homeland. The black species, created for hot and humid conditions, will prevail in the Southern lowlands, though whites will maintain dominion over the seashore and elevated ground. The new South will contain some Negro states. We should bow before this necessity and admit them into the Union; we have, after all, already recognized both "Haity and Liberia."* But the bracing North is not a congenial home for carefree and lackadaisical people, created for warmer regions. Pure blacks will migrate South, leaving a stubborn residue to dwindle and die out in the North: "I hope it may gradually die out in the north where it has only an artificial foothold" (11 August 1863). As for the mulattoes, "their sickly physique and their impaired fecundity" should assure their demise once the shackles of slavery no longer provide an opportunity for unnatural interbreeding.

Agassiz's world collapsed during the last decade of his life. His students rebelled; his supporters defected. He remained a hero to the public, but scientists began to regard him as a rigid and aging dogmatist, standing firm in his antiquated beliefs before the Darwinian tide. But his social preferences for racial segregation prevailed—all the more because his fanciful hope for voluntary geographic separation did not.

Samuel George Morton—empiricist of polygeny

Agassiz did not spend all his time in Philadelphia reviling black waiters. In the same letter to his mother, he wrote in glowing terms of his visit to the anatomical collection of Philadelphia's distinguished scientist and physician Samuel George Morton: "Imagine a series of 600 skulls, most of Indians from all tribes who inhabit or once inhabited all of America. Nothing like it exists anywhere else. This collection, by itself, is worth a trip to America" (Agassiz to his mother, December 1846, translated from the original letter in Houghton Library, Harvard University).

*Not all detractors of blacks were so generous. E. D. Cope, who feared that miscegenation would block the path to heaven (see preceding footnote), advocated the return of all blacks to Africa (1890, p. 2053): "Have we not burdens enough to carry in the European peasantry which we are called on every year to receive and assimilate? Is our own race on a plane sufficiently high, to render it safe for us to carry eight millions of dead material in the very center of our vital organism?"

Agassiz speculated freely and at length, but he amassed no data to support his polygenic theory. Morton, a Philadelphia patrician with two medical degrees—one from fashionable Edinburgh—provided the "facts" that won worldwide respect for the "American school" of polygeny. Morton began his collection of human skulls in the 1820s; he had more than one thousand when he died in 1851. Friends (and enemies) referred to his great charnel house as "the American Golgotha."

Morton won his reputation as the great data-gatherer and objectivist of American science, the man who would raise an immature enterprise from the mires of fanciful speculation. Oliver Wendell Holmes praised Morton for "the severe and cautious character" of his works, which "from their very nature are permanent data for all future students of ethnology" (in Stanton, 1960, p. 96). The same Humboldt who had asserted the inherent equality of all races wrote:

> The craniological treasures which you have been so fortunate as to unite in your collection, have in you found a worthy interpreter. Your work is equally remarkable for the profundity of its anatomical views, the numerical detail of the relations of organic conformation, and the absence of those poetical reveries which are the myths of modern physiology (in Meigs, 1851, p. 48).

When Morton died in 1851, the *New York Tribune* wrote that "probably no scientific man in America enjoyed a higher reputation among scholars throughout the world, than Dr. Morton" (in Stanton, 1960, p. 144).

Yet Morton gathered skulls neither for the dilettante's motive of abstract interest nor the taxonomist's zeal for complete representation. He had a hypothesis to test: that a ranking of races could be established objectively by physical characteristics of the brain, particularly by its size. Morton took a special interest in native Americans. As George Combe, his fervent friend and supporter, wrote:

> One of the most singular features in the history of this continent, is, that the aboriginal races, with few exceptions, have perished or constantly receded, before the Anglo-Saxon race, and have in no instance either mingled with them as equals, or adopted their manners and civilization. These phenomena must have a cause; and can any inquiry be at once more inter-

esting and philosophical than that which endeavors to ascertain whether that cause be connected with a difference in the brain between the native American race, and their conquering invaders (Combe and Coates, in review of Morton's *Crania Americana,* 1840, p. 352).

Moreover, Combe argued that Morton's collection would acquire true scientific value *only if* mental and moral worth could be read from brains: "If this doctrine be unfounded, these skulls are mere facts in Natural History, presenting no particular information as to the mental qualities of the people" (from Combe's appendix to Morton's *Crania Americana,* 1839, p. 275).

Although he vacillated early in his career, Morton soon became a leader among the American polygenists. He wrote several articles to defend the status of human races as separate, created species. He took on the strongest claim of opponents—the interfertility of all human races—by arguing from both sides. He relied on travelers' reports to claim that some human races—Australian aborigines and Caucasians in particular—very rarely produce fertile offspring (Morton, 1851). He attributed this failure to "a disparity of primordial organization." But, he continued, Buffon's criterion of interfertility must be abandoned in any case, for hybridization is common in nature, even between species belonging to different genera (Morton, 1847, 1850). Species must be redefined as "a primordial organic form" (1850, p. 82). "Bravo, my dear Sir," wrote Agassiz in a letter, "you have at last furnished science with a true philosophical definition of species" (in Stanton, 1960, p. 141). But how to recognize a primordial form? Morton replied: "If certain existing organic types can be traced back into the 'night of time,' as dissimilar as we see them now, is it not more reasonable to regard them as aboriginal, than to suppose them the mere and accidental derivations of an isolated patriarchal stem of which we know nothing?" (1850, p. 82). Thus, Morton regarded several breeds of dogs as separate species because their skeletons resided in the Egyptian catacombs, as recognizable and distinct from other breeds as they are now. The tombs also contained blacks and Caucasians. Morton dated the beaching of Noah's Ark on Ararat at 4,179 years before his time, and the Egyptian tombs at just 1,000 years after that—clearly not enough time for the sons of Noah to differentiate into races. (How, he asks, can we believe that races changed so rapidly for 1,000 years, and not at all for 3,000 years since then?) Human races must have been separate from the start (Morton, 1839, p. 88).

But separate, as the Supreme Court once said, need not mean unequal. Morton therefore set out to establish relative rank on "objective" grounds. He surveyed the drawings of ancient Egypt and found that blacks are invariably depicted as menials—a sure sign that they have always played their appropriate biological role: "Negroes were numerous in Egypt, but their social position in ancient times was the same that it is now, that of servants and slaves" (Morton, 1844, p. 158). (A curious argument, to be sure, for these blacks had been captured in warfare; sub-Saharan societies depicted blacks as rulers.)

But Morton's fame as a scientist rested upon his collection of skulls and their role in racial ranking. Since the cranial cavity of a human skull provides a faithful measure of the brain it once contained, Morton set out to rank races by the average sizes of their brains. He filled the cranial cavity with sifted white mustard seed, poured the seed back into a graduated cylinder and read the skull's volume in cubic inches. Later on, he became dissatisfied with mustard seed because he could not obtain consistent results. The seeds did not pack well, for they were too light and still varied too much in size, despite sieving. Remeasurements of single skulls might differ by more than 5 percent, or 4 cubic inches in skulls with an average capacity near 80 cubic inches. Consequently, he switched to one-eighth-inch-diameter lead shot "of the size called BB" and achieved consistent results that never varied by more than a single cubic inch for the same skull.

Morton published three major works on the sizes of human skulls—his lavish, beautifully illustrated volume on American Indians, the *Crania Americana* of 1839; his studies on skulls from the Egyptian tombs, the *Crania Aegyptiaca* of 1844; and the epitome of his entire collection in 1849. Each contained a table, summarizing his results on average skull volumes arranged by race. I have reproduced all three tables here (Tables 2.1 to 2.3). They represent the major contribution of American polygeny to debates about racial ranking. They outlived the theory of separate creations and were reprinted repeatedly during the nineteenth century as irrefutable, "hard" data on the mental worth of human races (see p. 84). Needless to say, they matched every good Yankee's prejudice—whites on top, Indians in the middle, and blacks on the bot-

Table 2 • 1 *Morton's summary table of cranial capacity by race*

RACE	N	INTERNAL CAPACITY (IN^3) MEAN	LARGEST	SMALLEST
Caucasian	52	87	109	75
Mongolian	10	83	93	69
Malay	18	81	89	64
American	147	82	100	60
Ethiopian	29	78	94	65

Table 2 • 2 *Cranial capacities for skulls from Egyptian tombs*

PEOPLE	MEAN CAPACITY (IN^3)	N
Caucasian		
Pelasgic	88	21
Semitic	82	5
Egyptian	80	39
Negroid	79	6
Negro	73	1

tom; and, among whites, Teutons and Anglo-Saxons on top, Jews in the middle, and Hindus on the bottom. Moreover, the pattern had been stable throughout recorded history, for whites had the same advantage over blacks in ancient Egypt. Status and access to power in Morton's America faithfully reflected biological merit. How could sentimentalists and egalitarians stand against the dictates of nature? Morton had provided clean, objective data based on the largest collection of skulls in the world.

During the summer of 1977 I spent several weeks reanalyzing Morton's data. (Morton, the self-styled objectivist, published all his raw information. We can infer with little doubt how he moved from raw measurements to summary tables.) In short, and to put it bluntly, Morton's summaries are a patchwork of fudging and finagling in the clear interest of controlling a priori convictions. Yet—and this is the most intriguing aspect of the case—I find no evidence of conscious fraud; indeed, had Morton been a conscious fudger, he would not have published his data so openly.

Conscious fraud is probably rare in science. It is also not very interesting, for it tells us little about the nature of scientific activity.

Table 2 • 3 *Morton's final summary of cranial capacity by race*

RACES AND FAMILIES	N	CRANIAL CAPACITY (IN^3) LARGEST	SMALLEST	MEAN	MEAN
MODERN CAUCASIAN GROUP					
Teutonic Family					
Germans	18	114	70	90	92
English	5	105	91	96	
Anglo-Americans	7	97	82	90	
Pelasgic Family	10	94	75	84	
Celtic Family	6	97	78	87	
Indostanic Family	32	91	67	80	
Semitic Family	3	98	84	89	
Nilotic Family	17	96	66	80	
ANCIENT CAUCASIAN GROUP					
Pelasgic Family	18	97	74	88	
Nilotic Family	55	96	68	80	
MONGOLIAN GROUP					
Chinese Family	6	91	70	82	
MALAY GROUP					
Malayan Family	20	97	68	86	85
Polynesian Family	3	84	82	83	
AMERICAN GROUP					
Toltecan Family					
Peruvians	155	101	58	75	79
Mexicans	22	92	67	79	
Barbarous Tribes	161	104	70	84	
NEGRO GROUP					
Native African Family	62	99	65	83	83
American-born Negroes	12	89	73	82	
Hottentot Family	3	83	68	75	
Australians	8	83	63	75	

Liars, if discovered, are excommunicated; scientists declare that their profession has properly policed itself, and they return to work, mythology unimpaired, and objectively vindicated. The prevalence of *unconscious* finagling, on the other hand, suggests a general conclusion about the social context of science. For if scientists can be honestly self-deluded to Morton's extent, then prior prejudice may be found anywhere, even in the basics of measuring bones and toting sums.

The case of Indian inferiority: Crania Americana*

Morton began his first and largest work, the *Crania Americana* of 1839, with a discourse on the essential character of human races. His statements immediately expose his prejudices. Of the "Greenland esquimaux," he wrote: "They are crafty, sensual, ungrateful, obstinate and unfeeling, and much of their affection for their children may be traced to purely selfish motives. They devour the most disgusting aliments uncooked and uncleaned, and seem to have no ideas beyond providing for the present moment. . . . Their mental faculties, from infancy to old age, present a continued childhood. . . . In gluttony, selfishness and ingratitude, they are perhaps unequalled by any other nation of people" (1839, p. 54). Morton thought little better of other Mongolians, for he wrote of the Chinese (p. 50): "So versatile are their feelings and actions, that they have been compared to the monkey race, whose attention is perpetually changing from one object to another." The Hottentots, he claimed (p. 90), are "the nearest approximation to the lower animals. . . . Their complexion is a yellowish brown, compared by travellers to the peculiar hue of Europeans in the last stages of jaundice. . . . The women are represented as even more repulsive in appearance than the men." Yet, when Morton had to describe one Caucasian tribe as a "mere horde of rapacious banditti" (p. 9), he quickly added that "their moral perceptions, under the influence of an equitable government, would no doubt assume a much more favorable aspect."

Morton's summary chart (Table 2.1) presents the "hard" argument of the *Crania Americana.* He had measured the capacity of 144 Indian skulls and calculated a mean of 82 cubic inches, a full 5 cubic inches below the Caucasian norm (Figs. 2.4 and 2.5). In addition, Morton appended a table of phrenological measurements indicating a deficiency of "higher" mental powers among Indians. "The benevolent mind," Morton concluded (p. 82), "may regret

*This account omits many statistical details of my analysis. The complete tale appears in Gould, 1978. Some passages in pp. 56–69 are taken from this article.

the inaptitude of the Indian for civilization," but sentimentality must yield to fact. "The structure of his mind appears to be different from that of the white man, nor can the two harmonize in the social relations except on the most limited scale." Indians "are not only averse to the restraints of education, but for the most part are incapable of a continued process of reasoning on abstract subjects" (p. 81).

Since *Crania Americana* is primarily a treatise on the inferior quality of Indian intellect, I note first of all that Morton's cited average of 82 cubic inches for Indian skulls is incorrect. He separated Indians into two groups, "Toltecans" from Mexico and South America, and "Barbarous Tribes" from North America. Eighty-two is the average for Barbarous skulls; the total sample of 144 yields a mean of 80.2 cubic inches, or a gap of almost 7 cubic inches between Indian and Caucasian averages. (I do not know how Morton made this elementary error. It did permit him, in any case, to retain the conventional chain of being with whites on top, Indians in the middle, and blacks on the bottom.)

But the "correct" value of 80.2 is far too low, for it is the result of an improper procedure. Morton's 144 skulls belong to many different groups of Indians; these groups differ significantly among themselves in cranial capacity. Each group should be weighted equally, lest the final average be biased by unequal size of subsamples. Suppose, for example, that we tried to estimate average human height from a sample of two jockeys, the author of this book (strictly middling stature), and all the players in the National Basketball Association. The hundreds of Jabbars would swamp the remaining three and give an average in excess of six and a half feet. If, however, we averaged the averages of the three groups (jockeys, me, and the basketball players), then our figure would lie closer to the true value. Morton's sample is strongly biased by a major overrepresentation of an extreme group—the small-brained Inca Peruvians. (They have a mean cranial capacity of 74.36 cubic inches and provide 25 percent of the entire sample). Large-brained Iroquois, on the other hand, contribute only 3 skulls to the total sample (2 percent). If, by the accidents of collecting, Morton's sample had included 25 percent Iroquois and just a few Incas, his average would have risen substantially. Consequently, I corrected this bias as best I could by averaging the mean values for all tribes

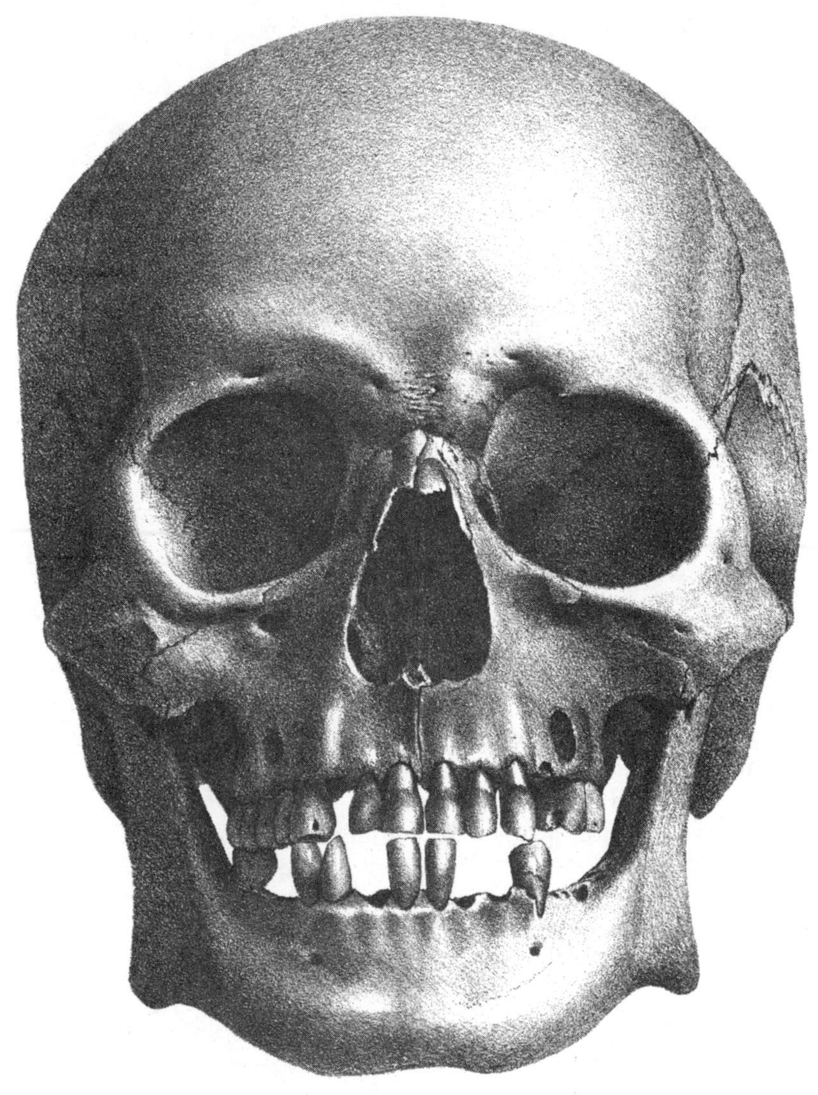

2•4 The skull of an Araucanian Indian. The lithographs of this and the next figure were done by John Collins, a great scientific artist unfortunately unrecognized today. They appeared in Morton's *Crania Americana* of 1839.

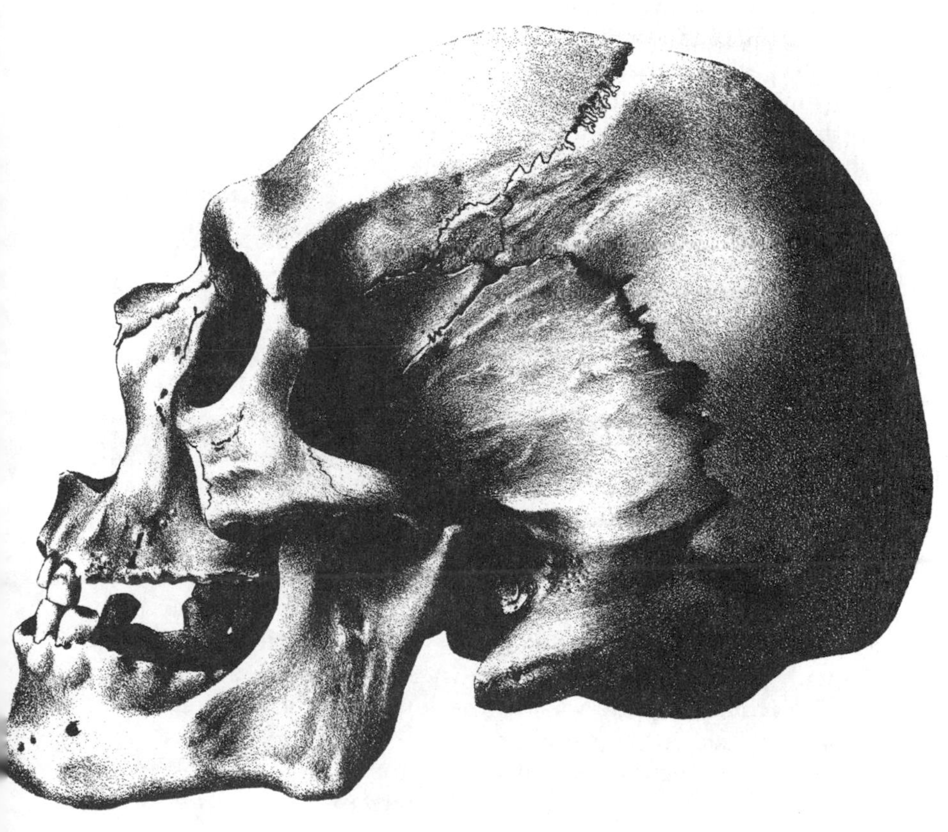

2•5 The skull of a Huron Indian. Lithograph by John Collins from Morton's *Crania Americana,* 1839.

represented by 4 or more skulls. The Indian average now rises to 83.79 cubic inches.

This revised value is still more than 3 cubic inches from the Caucasian average. Yet, when we examine Morton's procedure for computing the Caucasian mean, we uncover an astounding inconsistency. Since statistical reasoning is largely a product of the last one hundred years, I might have excused Morton's error for the Indian mean by arguing that he did not recognize the biases produced by unequal sizes among subsamples. But now we discover that he understood this bias perfectly well—for Morton calculated his high Caucasian mean by consciously eliminating small-brained Hindus from his sample. He writes (p. 261): "It is proper, however, to mention that but 3 Hindoos are admitted in the whole number, because the skulls of these people are probably smaller than those of any other existing nation. For example, 17 Hindoo heads give a mean of but 75 cubic inches; and the three received into the table are taken at that average." Thus, Morton included a large subsample of small-brained people (Inca Peruvians) to pull down the Indian average, but excluded just as many small Caucasian skulls to raise the mean of his own group. Since he tells us what he did so baldly, we must assume that Morton did not deem his procedure improper. But by what rationale did he keep Incas and exclude Hindus, unless it were the a priori assumption of a truly higher Caucasian mean? For one might then throw out the Hindu sample as truly anomalous, but retain the Inca sample (with the same mean as the Hindus, by the way) as the lower end of normality for its disadvantaged larger group.

I restored the Hindu skulls to Morton's sample, using the same procedure of equal weighting for all groups. Morton's Caucasian sample, by his reckoning, contains skulls from four subgroups, so Hindus should contribute one-fourth of all skulls to the sample. If we restore all seventeen of Morton's Hindu skulls, they form 26 percent of the total sample of sixty-six. The Caucasian mean now drops to 84.45 cubic inches, for no difference worth mentioning between Indians and Caucasians. (Eskimos, despite Morton's low opinion of them, yield a mean of 86.8, hidden by amalgamation with other subgroups in the Mongol grand mean of 83). So much for Indian inferiority.

The case of the Egyptian catacombs: Crania Aegyptiaca

Morton's friend and fellow polygenist George Gliddon was United States consul for the city of Cairo. He dispatched to Philadelphia more than one hundred skulls from tombs of ancient Egypt, and Morton responded with his second major treatise, the *Crania Aegyptiaca* of 1844. Morton had shown, or so he thought, that whites surpassed Indians in mental endowment. Now he would crown his story by demonstrating that the discrepancy between whites and blacks was even greater, and that this difference had been stable for more than three thousand years.

Morton felt that he could identify both races and subgroups among races from features of the skull (most anthropologists today would deny that such assignments can be made unambiguously). He divided his Caucasian skulls into Pelasgics (Hellenes, or ancient Greek forebears), Jews, and Egyptians—in that order, again confirming Anglo-Saxon preferences (Table 2.2). Non-Caucasian skulls he identified either as "negroid" (hybrids of Negro and Caucasian with more black than white) or as pure Negro.

Morton's subjective division of Caucasian skulls is clearly unwarranted, for he simply assigned the most bulbous crania to his favored Pelasgic group and the most flattened to Egyptians; he mentions no other criteria of subdivision. If we ignore his threefold separation and amalgamate all sixty-five Caucasian skulls into a single sample, we obtain an average capacity of 82.15 cubic inches. (If we give Morton the benefit of all doubt and rank his dubious subsamples equally—as we did in computing Indian and Caucasian means for the *Crania Americana*—we obtain an average of 83.3 cubic inches.)

Either of these values still exceeds the negroid and Negro averages substantially. Morton assumed that he had measured an innate difference in intelligence. He never considered any other proposal for the disparity in average cranial capacity—though another simple and obvious explanation lay before him.

Sizes of brains are related to the sizes of bodies that carry them: big people tend to have larger brains than small people. This fact does not imply that big people are smarter—any more than elephants should be judged more intelligent than humans because their brains are larger. Appropriate corrections must be made for differences in body size. Men tend to be larger than women; consequently, their brains are bigger. When corrections for body size are applied, men and women have brains of approximately equal size. Morton not only failed to correct for differences in sex or body size; he did not even recognize the relationship, though his data proclaimed it loud and clear. (I can only conjecture that Morton never separated his skulls by sex or stature—though his tables record these data—because he wanted so much to read differences in brain size directly as differences in intelligence.)

Many of the Egyptian skulls came with mummified remains of their possessors (Fig. 2.6), and Morton could record their sex unambiguously. If we use Morton's own designations and compute separate averages for males and females (as Morton never did), we obtain the following remarkable result. Mean capacity for twenty-four Caucasian skulls is 86.5 cubic inches; twenty-two female skulls average 77.2 (the remaining nineteen skulls could not be identified by sex). Of the six negroid skulls, Morton identified two as female (at 71 and 77 cubic inches) and could not allocate the other four (at 77, 77, 87, and 88).* If we make the reasonable conjecture that the two smaller skulls (77 and 77) are female, and the two larger male (87 and 88), we obtain a male negroid average of 87.5, slightly above the Caucasian male mean of 86.5, and a female negroid average of 75.5, slightly below the Caucasian value of 77.2. The apparent difference of 4 cubic inches between Morton's Caucasian and negroid samples may only record the fact that about half his Caucasian sample is male, while only one-third the negroid sample may be male. (The apparent difference is magnified by Morton's incorrect rounding of the negroid average down to 79 rather than up to 80. As we shall see again, all of Morton's minor numerical errors favor his prejudices.) Differences in average brain size between Caucasians and negroids in the Egyptian tombs only record differences in stature due to sex, not variation in "intelligence." You will not be surprised to learn that the single pure Negro skull (73 cubic inches) is a female.

*In his final catalogue of 1849, Morton guessed at sex (and age within five years!) for all crania. In this later work, he specifies 77, 87, and 88 as male, and the remaining 77 as female. This allocation was pure guesswork; my alternate version is equally plausible. In the *Crania Aegyptiaca* itself, Morton was more cautious and only identified sex for specimens with mummified remains.

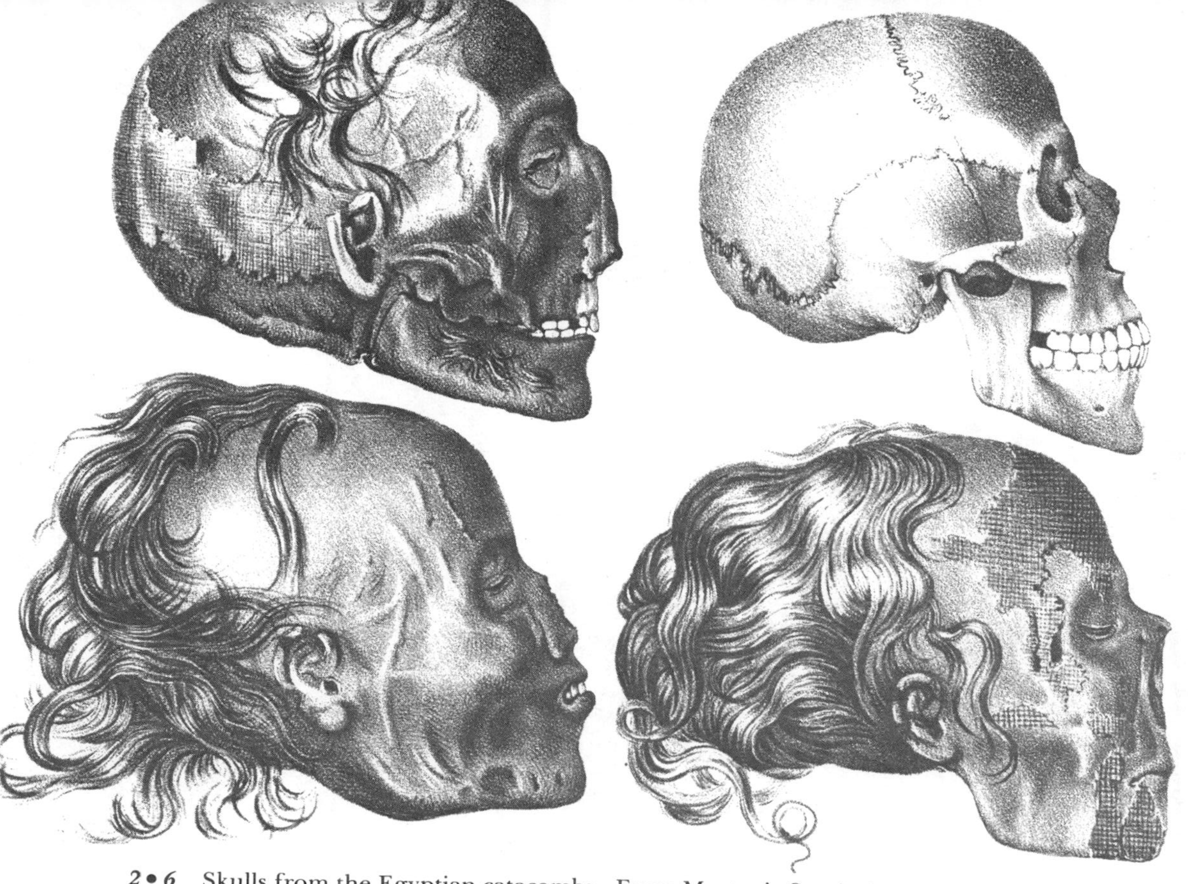

2•6 Skulls from the Egyptian catacombs. From Morton's *Crania Aegyptiaca* of 1844.

Table 2 • 4 *Cranial capacity of Indian groups ordered by Morton's assessment of body stature*

STATURE AND GROUP	CRANIAL CAPACITY (IN^3)	N
LARGE		
Seminole-Muskogee	88.3	8
Chippeway and related groups	88.8	4
Dacota and Osage	84.4	7
MIDDLE		
Mexicans	80.2	13
Menominee	80.5	8
Mounds	81.7	9
SMALL		
Columbia River Flatheads	78.8	10
Peruvians	74.4	33

The correlation of brain and body also resolves a question left hanging in our previous discussion of the *Crania Americana:* What is the basis for differences in average brain size among Indian peoples? (These differences bothered Morton considerably, for he could not understand how small-brained Incas had built such an elaborate civilization, though he consoled himself with the fact of their rapid conquest by the conquistadores). Again, the answer lay before him, but Morton never saw it. Morton presents subjective data on bodily statures in his descriptions of the various tribes, and I present these assessments along with average brain sizes in Table 2.4. The correlation of brain and body size is affirmed without exception. The low Hindu mean among Caucasians also records a difference in stature, not another case of dumb Indians.

The case of the shifting black mean

In the *Crania Americana,* Morton cited 78 cubic inches as the average cranial capacity for blacks. Five years later, in the *Crania Aegyptiaca,* he appended the following footnote to his table of measurements: "I have in my possession 79 crania of Negroes born in Africa. . . . Of the whole number, 58 are adult . . . and give 85 cubic inches for the average size of the brain" (1844, p. 113).

Since Morton had changed his method of measurement from

mustard seed to lead shot between 1839 and 1844, I suspected this alteration as a cause for the rising black mean. Fortunately, Morton remeasured most of his skulls personally, and his various catalogues present tabulations of the same skulls by both seed and shot (see Gould, 1978, for details).

I assumed that measures by seed would be lower. Seeds are light and variable in size, even after sieving. Hence, they do not pack well. By vigorous shaking or pressing of the thumb at the foramen magnum (the hole at the base of a skull), seeds can be made to settle, providing room for more. Measures by seed were very variable; Morton reported differences of several cubic inches for recalibrations of the same skull. He eventually became discouraged, fired his assistants, and redid all his measurements personally, with lead shot. Recalibrations never varied by more than a cubic inch, and we may accept Morton's judgment that measures by shot were objective, accurate, and repeatable—while earlier measures by seed were highly subjective and erratic.

I then calculated the discrepancies between seed and shot by race. Shot, as I suspected, always yielded higher values than seed. For 111 Indian skulls, measured by both criteria, shot exceeds seed by an average of 2.2 cubic inches. Data are not as reliable for blacks and Caucasians because Morton did not specify individual skulls for these races in the *Crania Americana* (measured by seed). For Caucasians, 19 identifiable skulls yield an average discrepancy of only 1.8 cubic inches for shot over seed. Yet 18 African skulls, remeasured from the sample reported in *Crania Americana*, produce a mean by shot of 83.44 cubic inches, a rise of 5.4 cubic inches from the 1839 average by seed. In other words, the more "inferior" a race by Morton's a priori judgment, the greater the discrepancy between a subjective measurement, easily and unconsciously fudged, and an objective measure unaffected by prior prejudice. The discrepancy for blacks, Indians, and Caucasians is 5.4, 2.2, and 1.8 cubic inches, respectively.

Plausible scenarios are easy to construct. Morton, measuring by seed, picks up a threateningly large black skull, fills it lightly and gives it a few desultory shakes. Next, he takes a distressingly small Caucasian skull, shakes hard, and pushes mightily at the foramen magnum with his thumb. It is easily done, without conscious motivation; expectation is a powerful guide to action.

Table 2 • 5 *Corrected values for Morton's final tabulation*

PEOPLE	CRANIAL CAPACITY (IN^3)
Mongolians	87
Modern Caucasians	87
Native Americans	86
Malays	85
Ancient Caucasians	84
Africans	83

The final tabulation of 1849

Morton's burgeoning collection included 623 skulls when he presented his final tabulation in 1849—an overwhelming affirmation of the ranking that every Anglo-Saxon expected.

The Caucasian subsamples suffer from errors and distortions. The German mean, reported at 90 in the summary, is 88.4 from individual skulls listed in the catalogue; the correct Anglo-American average is 89 (89.14), not 90. The high English mean of 96 is correct, but the small sample is entirely male.* If we follow our procedure of computing averages among subsamples, the six modern Caucasian "families" yield a mean of 87 cubic inches.† The ancient Caucasian average for two subsamples is 84 cubic inches (Table 2.5).

Six Chinese skulls provide Morton with a Mongolian mean of 82, but this low value records two cases of selective amnesia: First,

*To demonstrate again how large differences based on stature can be, I report these additional data, recovered from Morton's tabulations, but never calculated or recognized by him: 1) For Inca Peruvians, fifty-three male skulls average 77.5; sixty-one female skulls, 72.1. 2) For Germans, nine male skulls average 92.2; eight females, 84.3.

†My original report (Gould, 1978) incorrectly listed the modern Caucasian mean as 85.3. The reason for this error is embarrassing, but instructive, for it illustrates, at my expense, the cardinal principle of this book: the social embeddedness of science and the frequent grafting of expectation upon supposed objectivity. Line 7 in Table 2.3 lists the range of Semitic skulls as 84 to 98 cubic inches for Morton's sample of 3. However, my original paper cited a mean of 80—an obvious impossibility if the smallest skull measures 84. I was working from a Xerox of Morton's original chart, and his correct value of 89 is smudged to look like an 80 on my copy. Nonetheless, the range of 84 to 98 is clearly indicated right alongside, and I never saw the inconsistency—presumably because a low value of 80 fit my hopes for a depressed Caucasian mean. The 80 therefore "felt" right and I never checked it. I am grateful to Dr. Irving Klotz of Northwestern University for pointing out this error to me.

Morton excluded the latest Chinese specimen (skull number 1336 at 98 cubic inches), though it must have been in his collection when he published his summary because he includes many Peruvian skulls with higher numbers. Secondly, although Morton deplored the absence of Eskimos from his collection (1849, p. iv), he did not mention the three Eskimo skulls that he had measured for *Crania Americana*. (These belonged to his friend George Combe and do not appear in Morton's final catalogue.)

Morton never remeasured these skulls with shot, but if we apply the Indian correction of 2.2 cubic inches to their seed average of 86.8 we obtain a mean of 89. These two samples (Chinese with number 1336 added, and Eskimo conservatively corrected) yield a Mongolian average of 87 cubic inches.

By 1849 Morton's Indian mean had plummeted to 79. But this figure is invalid for the same reason as before, though now intensified—inequality of numbers among subsamples. Small-headed (and small-statured) Peruvians provided 23 percent of the 1839 sample, but their frequency had risen to nearly half (155 of 338 skulls) by 1849. If we use our previous criterion and compute the average of all subsamples weighted equally, the Indian average is 86 cubic inches.

For the Negro average, we should drop Morton's australoids because he wanted to assess the status of African blacks and we no longer accept a close relationship between the two groups—dark skin evolved more than once among human groups. I also drop the Hottentot sample of 3. All skulls are female, and Hottentots are very small in stature. Native and American-born blacks, amalgamated to a single sample, yield an average value between 82 and 83, but closer to 83.

In short, my correction of Morton's conventional ranking reveals *no* significant differences among races for Morton's own data (Table 2.5). All groups rank between 83 and 87 cubic inches, and Caucasians share the pinnacle. If western Europeans choose to seek their superiority in high averages for their subsamples (Germanics and Anglo-Saxons in the Caucasian tabulations), I point out that several Indian subsamples are equally high (though Morton amalgamated all North American Indians and never reported averages by subgroup), and that all Teutonic and Anglo-Saxon averages are either miscalculated or biased in Morton's table.

Conclusions

Morton's finagling may be ordered into four general categories:

1. Favorable inconsistencies and shifting criteria: Morton often chose to include or delete large subsamples in order to match group averages with prior expectations. He included Inca Peruvians to decrease the Indian average, but deleted Hindus to raise the Caucasian mean. He also chose to present or not to calculate the averages of subsamples in striking accord with desired results. He made calculations for Caucasians to demonstrate the superiority of Teutons and Anglo-Saxons, but never presented data for Indian subsamples with equally high averages.

2. Subjectivity directed toward prior prejudice: Morton's measures with seed were sufficiently imprecise to permit a wide range of influence by subjective bias; later measures with shot, on the other hand, were repeatable, and presumably objective. In skulls measured by both methods, values for shot always exceed values for the light, poorly packing seed. But degrees of discrepancy match a priori assumptions: an average of 5.4, 2.2, and 1.8 cubic inches for blacks, Indians, and whites, respectively. In other words, blacks fared poorest and whites best when the results could be biased toward an expected result.

3. Procedural omissions that seem obvious to us: Morton was convinced that variation in skull size recorded differential, innate mental ability. He never considered alternate hypotheses, though his own data almost cried out for a different interpretation. Morton never computed means by sex or stature, even when he recorded these data in his tabulations—as for Egyptian mummies. Had he computed the effect of stature, he would presumably have recognized that it explained all important differences in brain size among his groups. Negroids yielded a lower average than Caucasians among his Egyptian skulls because the negroid sample probably contained a higher percentage of smaller-statured females, not because blacks are innately stupider. The Incas that he included in the Indian sample and the Hindus that he excluded from the Caucasian sample both possessed small brains as a consequence of small body size. Morton used an all-female sample of three Hottentots to support the stupidity of blacks, and an all-male sample of Englishmen to assert the superiority of whites.

4. Miscalculations and convenient omissions: All miscalculations and omissions that I have detected are in Morton's favor. He rounded the negroid Egyptian average down to 79, rather than up to 80. He cited averages of 90 for Germans and Anglo-Saxons, but the correct values are 88 and 89. He excluded a large Chinese skull and an Eskimo subsample from his final tabulation for mongoloids, thus depressing their average below the Caucasian value.

Yet through all this juggling, I detect no sign of fraud or conscious manipulation. Morton made no attempt to cover his tracks and I must presume that he was unaware he had left them. He explained all his procedures and published all his raw data. All I can discern is an a priori conviction about racial ranking so powerful that it directed his tabulations along preestablished lines. Yet Morton was widely hailed as the objectivist of his age, the man who would rescue American science from the mire of unsupported speculation.

The American school and slavery

The leading American polygenists differed in their attitude toward slavery. Most were Northerners, and most favored some version of Squier's quip: "[I have a] precious poor opinion of niggers . . . a still poorer one of slavery" (in Stanton, 1960, p. 193).

But the identification of blacks as a separate and unequal species had obvious appeal as an argument for slavery. Josiah Nott, a leading polygenist, encountered particularly receptive audiences in the South for his "lectures on niggerology" (as he called them). Morton's *Crania Aegyptiaca* received a warm welcome in the South (in Stanton, 1960, pp. 52–53). One supporter of slavery wrote that the South need no longer be "so much frightened" by "voices of Europe or of Northern America" in defending its "peculiar institutions." When Morton died, the South's leading medical journal proclaimed (R. W. Gibbs, *Charleston Medical Journal,* 1851, quoted in Stanton, 1960, p. 144): "We of the South should consider him as our benefactor, for aiding most materially in giving to the negro his true position as an inferior race."

Nonetheless, the polygenist argument did not occupy a primary place in the ideology of slavery in mid-nineteenth-century America—and for a good reason. For most Southerners, this excellent argument entailed too high a price. The polygenists had railed against ideologues as barriers to their pure search for truth, but their targets were parsons more often than abolitionists. Their theory, in asserting a plurality of human creations, contradicted the doctrine of a single Adam and contravened the literal truth of scripture. Although the leading polygenists held a diversity of religious attitudes, none were atheists. Morton and Agassiz were conventionally devout, but they did believe that both science and religion would be aided if untrained parsons kept their noses out of scientific issues and stopped proferring the Bible as a document to settle debates in natural history. Josiah Nott stated his goal in a forceful way (Agassiz and Morton would not have put it so baldly): ". . . to cut loose the natural history of mankind from the Bible, and to place each upon its own foundation, where it may remain without collision or molestation" (in Stanton, 1960, p. 119).

The polygenists forced defenders of slavery into a quandary: Should they accept a strong argument from science at the cost of limiting religion's sphere? In resolving this dilemma, the Bible usually won. After all, scriptural arguments for supporting slavery were not wanting. Degeneration of blacks under the curse of Ham was an old and eminently functional standby. Moreover, polygeny was not the only quasi-scientific defense available.

John Bachman, for example, was a South Carolina parson and prominent naturalist. As a committed monogenist, he spent a good part of his scientific career attempting to refute polygeny. He also used monogenist principles to defend slavery:

> In intellectual power the African is an inferior variety of our species. His whole history affords evidence that he is incapable of self-government. Our child that we lead by the hand, and who looks to us for protection and support is still of our own blood notwithstanding his weakness and ignorance (in Stanton, 1960, p. 63).

Among nonpolygenist, "scientific" defenses of slavery, no arguments ever matched in absurdity the doctrines of S. A. Cartwright, a prominent Southern physician. (I do not cite these as typical and I doubt that many intelligent Southerners paid them much attention; I merely wish to illustrate an extreme within the range of "scientific" argument.) Cartwright traced the problems of black people to inadequate decarbonization of blood in the lungs (insufficient removal of carbon dioxide): "It is the defective . . . atmospherization of the blood, conjoined with a deficiency of cerebral matter in the cranium . . . that is the true cause of that debasement of mind, which has rendered the people of Africa unable to take care of themselves" (from Chorover, 1979; all quotes from Cartwright are taken from papers he presented to the 1851 meeting of the Louisiana Medical Association.)

Cartwright even had a name for it—*dysesthesia,* a disease of inadequate breathing. He described its symptoms in slaves: "When driven to labor . . . he performs the task assigned to him in a headlong and careless manner, treading down with his feet or cutting with his hoe the plants he is put to cultivate—breaking the tools he works with, and spoiling everything he touches." Ignorant Northerners attributed this beahvior to "the debasing influence of slavery," but Cartwright recognized it as the expression of a true disease. He identified insensibility to pain as another symptom: "When the unfortunate individual is subjected to punishment, he neither feels pain of any consequence . . . [nor] any unusual resentment more than stupid sulkiness. In some cases . . . there appears to be an almost total loss of feeling." Cartwright proposed the following cure:

> The liver, skin and kidneys should be stimulated to activity . . . to assist in decarbonizing the blood. The best means to stimulate the skin is, first, to have the patient well washed with warm water and soap; then to anoint it all over with oil, and to slap the oil in with a broad leather strap; then to put the patient to some hard kind of work in the open air and sunshine that will compel him to expand his lungs, as chopping wood, splitting rails, or sawing with the crosscut or whip saw.

Cartwright did not end his catalogue of diseases with dysesthesia. He wondered why slaves often tried to flee, and identified the cause as a mental disease called *drapetomania,* or the insane desire to run away. "Like children, they are constrained by unalterable physiological laws, to love those in authority over them. Hence, from a law of his nature, the negro can no more help loving a kind master, than the child can help loving her that gives it suck." For slaves afflicted with drapetomania, Cartwright proposed a behavioral cure: owners should avoid both extreme permissiveness and cruelty: "They have only to be kept in that state, and treated like children, to prevent and cure them from running away."

The defenders of slavery did not need polygeny. Religion still stood above science as a primary source for the rationalization of social order. But the American debate on polygeny may represent the last time that arguments in the scientific mode did not form a first line of defense for the status quo and the unalterable quality of human differences. The Civil War lay just around the corner, but so did 1859 and Darwin's *Origin of Species*. Subsequent arguments for slavery, colonialism, racial differences, class structures, and sex roles would go forth primarily under the banner of science.

THREE

Measuring Heads

Paul Broca and the Heyday of Craniology

No rational man, cognisant of the facts, believes that the average negro is the equal, still less the superior, of the average white man. And, if this be true, it is simply incredible that, when all his disabilities are removed, and our prognathous relative has a fair field and no favor, as well as no oppressor, he will be able to compete successfully with his bigger-brained and smaller-jawed rival, in a contest which is to be carried on by thoughts and not by bites. —T. H. HUXLEY

The allure of numbers

Introduction

Evolutionary theory swept away the creationist rug that had supported the intense debate between monogenists and polygenists, but it satisfied both sides by presenting an even better rationale for their shared racism. The monogenists continued to construct linear hierarchies of races according to mental and moral worth; the polygenists now admitted a common ancestry in the prehistoric mists, but affirmed that races had been separate long enough to evolve major inherited differences in talent and intelligence. As historian of anthropology George Stocking writes (1973, p. lxx): "The resulting intellectual tensions were resolved after 1859 by a comprehensive evolutionism which was at once monogenist and racist, which affirmed human unity even as it relegated the dark-skinned savage to a status very near the ape."

The second half of the nineteenth century was not only the era of evolution in anthropology. Another trend, equally irresistible, swept through the human sciences—the allure of numbers, the faith that rigorous measurement could guarantee irrefutable precision, and might mark the transition between subjective speculation and a true science as worthy as Newtonian physics. Evolution and quantification formed an unholy alliance; in a sense, their union forged the first powerful theory of "scientific" racism—if we define "science" as many do who misunderstand it most profoundly: as any claim apparently backed by copious numbers. Anthropologists had presented numbers before Darwin, but the crudity of Morton's analysis (Chapter 2) belies any claim to rigor. By the end of Darwin's century, standardized procedures and a developing body of statistical knowledge had generated a deluge of more truthworthy numerical data.

This chapter is the story of numbers once regarded as surpassing all others in importance—the data of craniometry, or measurement of the skull and its contents. The leaders of craniometry were not conscious political ideologues. They regarded themselves as servants of their numbers, apostles of objectivity. And they confirmed all the common prejudices of comfortable white males—that blacks, women, and poor people occupy their subordinate roles by the harsh dictates of nature.

Science is rooted in creative interpretation. Numbers suggest, constrain, and refute; they do not, by themselves, specify the content of scientific theories. Theories are built upon the interpretation of numbers, and interpreters are often trapped by their own rhetoric. They believe in their own objectivity, and fail to discern the prejudice that leads them to one interpretation among many consistent with their numbers. Paul Broca is now distant enough. We can stand back and show that he used numbers not to generate new theories but to illustrate a priori conclusions. Shall we believe that science is different today simply because we share the cultural context of most practicing scientists and mistake its influence for objective truth? Broca was an exemplary scientist; no one has ever surpassed him in meticulous care and accuracy of measurement. By what right, other than our own biases, can we identify his prejudice and hold that science now operates independently of culture and class?

Francis Galton—apostle of quantification

No man expressed his era's fascination with numbers so well as Darwin's celebrated cousin, Francis Galton (1822–1911). Independently wealthy, Galton had the rare freedom to devote his considerable energy and intelligence to his favorite subject of measurement. Galton, a pioneer of modern statistics, believed that, with sufficient labor and ingenuity, anything might be measured, and that measurement is the primary criterion of a scientific study. He even proposed and began to carry out a statistical inquiry into the efficacy of prayer! Galton coined the term "eugenics" in 1883 and advocated the regulation of marriage and family size according to hereditary endowment of parents.

Galton backed his faith in measurement with all the ingenuity of his idiosyncratic methods. He sought, for example, to construct a "beauty map" of the British Isles in the following manner (1909, pp. 315–316):

> Whenever I have occasion to classify the persons I meet into three classes, "good, medium, bad," I use a needle mounted as a pricker, wherewith to prick holes, unseen, in a piece of paper, torn rudely into a cross with a long leg. I use its upper end for "good," the cross arm for "medium," the lower end for "bad." The prick holes keep distinct, and are easily read off at leisure. The object, place, and date are written on the paper. I used this plan for my beauty data, classifying the girls I passed in streets or elsewhere as attractive, indifferent, or repellent. Of course this was a purely individual estimate, but it was consistent, judging from the conformity of different attempts in the same population. I found London to rank highest for beauty; Aberdeen lowest.

With good humor, he suggested the following method for quantifying boredom (1909, p. 278):

> Many mental processes admit of being roughly measured. For instance, the degree to which people are bored, by counting the number of their fidgets. I not infrequently tried this method at the meetings of the Royal Geographical Society, for even there dull memoirs are occasionally read. . . . The use of a watch attracts attention, so I reckon time by the number of my breathings, of which there are 15 in a minute. They are not counted mentally, but are punctuated by pressing with 15 fingers successively. The counting is reserved for the fidgets. These observations should be confined to persons of middle age. Children are rarely still, while elderly philosophers will sometimes remain rigid for minutes altogether.

Quantification was Galton's god, and a strong belief in the inheritance of nearly everything he could measure stood at the right hand. Galton believed that even the most socially embedded behaviors had strong innate components: "As many members of our House of Lords marry the daughters of millionaires," he wrote (1909, pp. 314–315), "it is quite conceivable that our Senate may in time become characterized by a more than common share of shrewd business capacity, possibly also by a lower standard of commercial probity than at present." Constantly seeking new and ingenious ways to measure the relative worth of peoples, he proposed to rate blacks and whites by studying the history of encounters between black chiefs and white travelers (1884, pp. 338–339):

> The latter, no doubt, bring with them the knowledge current in civilized lands, but that is an advantage of less importance than we are apt to suppose. A native chief has as good an education in the art of ruling men, as can be desired; he is continually exercised in personal government, and usually maintains his place by the ascendancy of his character shown every day over his subjects and rivals. A traveller in wild countries also fills, to a certain degree, the position of a commander, and has to confront native chiefs at every inhabited place. The result is familiar enough—the white traveller almost invariably holds his own in their presence. It is seldom that we hear of a white traveller meeting with a black chief whom he feels to be the better man.

Galton's major work on the inheritance of intelligence (*Hereditary Genius,* 1869) included anthropometry among its criteria, but his interest in measuring skulls and bodies peaked later when he established a laboratory at the International Exposition of 1884. There, for threepence, people moved through his assembly line of tests and measures, and received his assessment at the end. After the Exposition, he maintained the lab for six years at a London museum. The laboratory became famous and attracted many notables, including Gladstone:

> Mr. Gladstone was amusingly insistent about the size of his head, saying that hatters often told him that he had an Aberdeenshire head—"a fact which you may be sure I do not forget to tell my Scotch constituents." It was a beautifully shaped head, though rather low, but after all it was not so very large in circumference (1909, pp. 249–250).

Lest this be mistaken for the harmless musings of some dotty Victorian eccentric, I point out that Sir Francis was taken quite seriously as a leading intellect of his time. The American hereditarian Lewis Terman, the man most responsible for instituting IQ tests in America, retrospectively calculated Galton's IQ at above 200, but accorded only 135 to Darwin and a mere 100–110 to Copernicus (see pp. 183–188 on this ludicrous incident in the history of mental testing). Darwin, who approached hereditarian arguments with strong suspicion, wrote after reading *Hereditary Genius:* "You have made a convert of an opponent in one sense, for I have always maintained that, excepting fools, men did not differ much in intellect, only in zeal and hard work" (in Galton, 1909, p. 290). Galton responded: "The rejoinder that might be made to his remark about hard work, is that character, including the aptitude for work, is heritable like every other faculty."

A curtain-raiser with a moral: Numbers do not guarantee truth

In 1906, a Virginia physician, Robert Bennett Bean, published a long, technical article comparing the brains of American blacks and whites. With a kind of neurological green thumb, he found meaningful differences wherever he looked—meaningful, that is, in his favored sense of expressing black inferiority in hard numbers.

Bean took special pride in his data on the corpus callosum, a structure within the brain that contains fibers connecting the right and left hemispheres. Following a cardinal tenet of craniometry, that higher mental functions reside in the front of the brain and sensorimotor capacities toward the rear, Bean reasoned that he might rank races by the relative sizes of parts within the corpus callosum. So he measured the length of the genu, the front part of the corpus callosum, and compared it with the length of the splenium, the back part. He plotted genu vs. splenium (Fig. 3.1) and obtained, for a respectably large sample, virtually complete separation between black and white brains. Whites have a relatively large genu, hence more brain up front in the seat of intelligence. All the more remarkable, Bean exclaimed (1906, p. 390) because the genu contains fibers both for olfaction and for intelligence! Bean continued: We all know that blacks have a keener sense of smell than whites; hence we might have expected larger genus in blacks if intelligence did not differ substantially between races. Yet black genus are smaller despite their olfactory predominance; hence, blacks must really suffer from a paucity of intelligence.

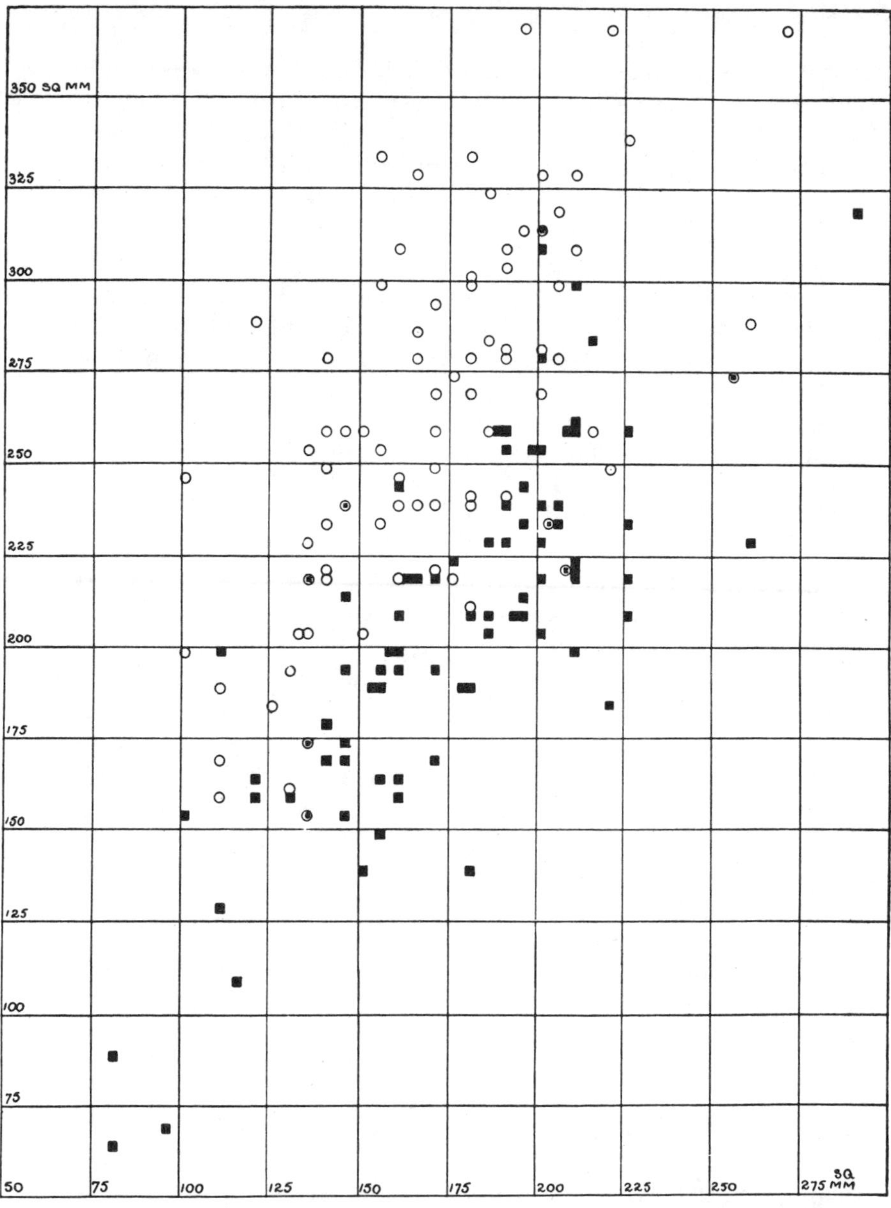

3 • 1 Bean's plot of the genu on the y-axis vs. the splenium on the x-axis. White circles are, unsurprisingly, for white brains; black squares for black brains. Whites seem to have a larger genu, hence more up front, and presumably more intelligence.

Moreover, Bean did not neglect to push the corresponding conclusion for sexes. Within each race, women have relatively smaller genus than men.

Bean then continued his discourse on the relatively greater size of frontal vs. parietal and occipital (side and back) parts of the brain in whites. In the relative size of their frontal areas, he proclaimed, blacks are intermediate between "man [*sic*] and the ourang-outang" (1906, p. 380).

Throughout this long monograph, one common measure is conspicuous by its absence: Bean says nothing about the size of the brain itself, the favored criterion of classical craniometry. The reason for this neglect lies buried in an addendum: black and white brains did not differ in overall size. Bean temporized: "So many factors enter into brain weight that it is questionable whether discussion of the subject is profitable here." Still, he found a way out. His brains came from unclaimed bodies given to medical schools. We all know that blacks have less respect for their dead than whites. Only the lowest classes of whites—prostitutes and the depraved—would be found among abandoned bodies, "while among Negroes it is known that even the better classes neglect their dead." Thus, even an absence of measured difference might indicate white superiority, for the data "do perhaps show that the low class Caucasian has a larger brain than a better class Negro" (1906, p. 409).

Bean's general conclusion, expressed in a summary paragraph before the troublesome addendum, proclaimed a common prejudice as the conclusion of science:

> The Negro is primarily affectionate, immensely emotional, then sensual and under stimulation passionate. There is love of ostentation, and capacity for melodious articulation; there is undeveloped artistic power and taste—Negroes make good artisans, handicraftsmen—and there is instability of character incident to lack of self-control, especially in connection with the sexual relation; and there is lack of orientation, or recognition of position and condition of self and environment, evidenced by a peculiar bumptiousness, so called, that is particularly noticeable. One would naturally expect some such character for the Negro, because the whole posterior part of the brain is large, and the whole anterior portion is small.

Bean did not confine his opinions to technical journals. He published two articles in popular magazines during 1906, and attracted sufficient attention to become the subject of an editorial in *American Medicine* for April 1907 (cited in Chase, 1977, p. 179). Bean had provided, the editorial proclaimed, "the anatomical basis for the complete failure of the negro schools to impart the higher studies—the brain cannot comprehend them any more than a horse can understand the rule of three. . . . Leaders in all political parties now acknowledge the error of human equality. . . . It may be practicable to rectify the error and remove a menace to our prosperity—a large electorate without brains."

But Franklin P. Mall, Bean's mentor at Johns Hopkins, became suspicious: Bean's data were too good. He repeated Bean's work, but with an important difference in procedure—he made sure that he did not know which brains were from blacks and which from whites until *after* he had measured them (Mall, 1909). For a sample of 106 brains, using Bean's method of measurement, he found no difference between whites and blacks in the relative sizes of genu and splenium (Fig. 3.2). This sample included 18 brains from Bean's original sample, 10 from whites, 8 from blacks. Bean's measure of the genu was larger than Moll's for 7 whites, but for only a single black. Bean's measure of the splenium was larger than Moll's for 7 of the 8 blacks.

I use this small tale of zealotry as a curtain-raiser because it illustrates so well the major contentions of this chapter and book:

1. Scientific racists and sexists often confine their label of inferiority to a single disadvantaged group; but race, sex, and class go together, and each acts as a surrogate for the others. Individual studies may be limited in scope, but the general philosophy of biological determinism pervades—hierarchies of advantage and disadvantage follow the dictates of nature; stratification reflects biology. Bean studied races, but he extended his most important conclusion to women, and also invoked differences of social class to argue that equality of size between black and white brains really reflects the inferiority of blacks.

2. Prior prejudice, not copious numerical documentation, dictates conclusions. We can scarcely doubt that Bean's statement about black bumptiousness reflected a prior belief that he set out to objectify, not an induction from data about fronts and backs of brains. And the special pleading that yielded black inferiority from equality of brain size is ludicrous outside a shared context of a priori belief in the inferiority of blacks.

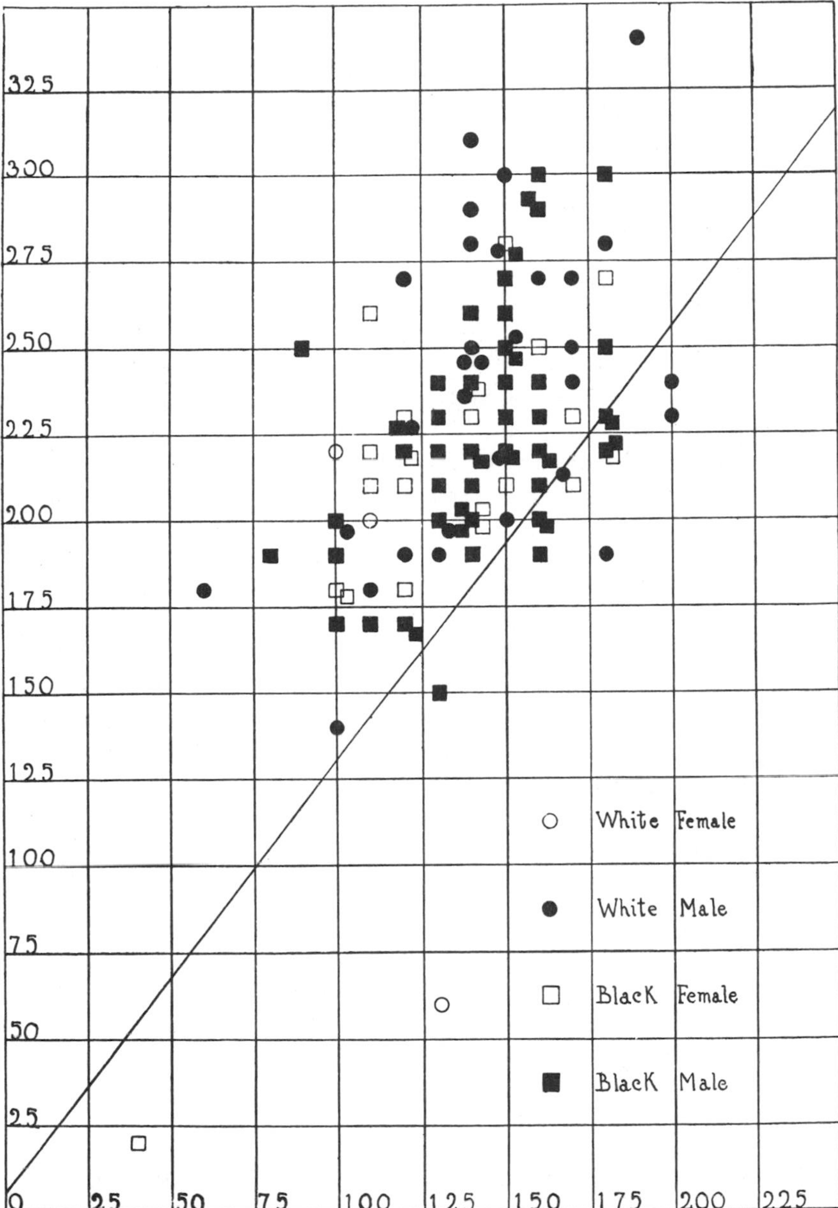

3 • 2 Moll's plot of genu vs. splenium. Moll measured the brains without knowing whether they came from whites or blacks. He found no difference between the races. The line represents Bean's separation between whites and blacks.

3. Numbers and graphs do not gain authority from increasing precision of measurement, sample size, or complexity in manipulation. Basic experimental designs may be flawed and not subject to correction by extended repetition. Prior commitment to one among many potential conclusions often guarantees a serious flaw in design.

4. Craniometry was not just a plaything of academicians, a subject confined to technical journals. Conclusions flooded the popular press. Once entrenched, they often embarked on a life of their own, endlessly copied from secondary source to secondary source, refractory to disproof because no one examined the fragility of primary documentation. In this case, Mall nipped a dogma in the bud, but not before a leading journal had recommended that blacks be barred from voting as a consequence of their innate stupidity.

But I also note an important difference between Bean and the great European craniometricians. Bean committed either conscious fraud or extraordinary self-delusion. He was a poor scientist following an absurd experimental design. The great craniometricians, on the other hand, were fine scientists by the criteria of their time. Their numbers, unlike Bean's, were generally sound. Their prejudices played a more subtle role in specifying interpretations and in suggesting what numbers might be gathered in the first place. Their work was more refractory to exposure, but equally invalid for the same reason: prejudices led through data in a circle back to the same prejudices—an unbeatable system that gained authority because it seemed to arise from meticulous measurement.

Bean's story has been told several times (Myrdal, 1944; Haller, 1971; Chase, 1977), if not with all its details. But Bean was a marginal figure on a temporary and provincial stage. I have found no modern analysis of the main drama, the data of Paul Broca and his school.

Masters of craniometry: Paul Broca and his school

The great circle route

In 1861 a fierce debate extended over several meetings of a young association still experiencing its birth pangs. Paul Broca

(1824–1880), professor of clinical surgery in the faculty of medicine, had founded the Anthropological Society of Paris in 1859. At a meeting of the society two years later, Louis Pierre Gratiolet read a paper that challenged Broca's most precious belief: Gratiolet dared to argue that the size of a brain bore no relationship to its degree of intelligence.

Broca rose in his own defense, arguing that "the study of the brains of human races would lose most of its interest and utility" if variation in size counted for nothing (1861, p. 141). Why had anthropologists spent so much time measuring skulls, unless their results could delineate human groups and assess their relative worth?

> Among the questions heretofore discussed within the Anthropological Society, none is equal in interest and importance to the question before us now. . . . The great importance of craniology has struck anthropologists with such force that many among us have neglected the other parts of our science in order to devote ourselves almost exclusively to the study of skulls. . . . In such data, we hoped to find some information relevant to the intellectual value of the various human races (1861, p. 139).

Broca then unleashed his data and poor Gratiolet was routed. His final contribution to the debate must rank among the most oblique, yet abject concession speeches ever offered by a scientist. He did not abjure his errors; he argued instead that no one had appreciated the subtlety of his position. (Gratiolet, by the way, was a royalist, not an egalitarian. He merely sought other measures to affirm the inferiority of blacks and women—earlier closure of the skull sutures, for example.)

Broca concluded triumphantly:

> In general, the brain is larger in mature adults than in the elderly, in men than in women, in eminent men than in men of mediocre talent, in superior races than in inferior races (1861, p. 304). . . . Other things equal, there is a remarkable relationship between the development of intelligence and the volume of the brain (p. 188).

Five years later, in an encyclopedia article on anthropology, Broca expressed himself more forcefully:

> A prognathous [forward-jutting] face, more or less black color of the skin, woolly hair and intellectual and social inferiority are often associated,

while more or less white skin, straight hair and an orthognathous [straight] face are the ordinary equipment of the highest groups in the human series (1866, p. 280). . . . A group with black skin, woolly hair and a prognathous face has never been able to raise itself spontaneously to civilization (pp. 295–296).

These are harsh words, and Broca himself regretted that nature had fashioned such a system (1866, p. 296). But what could he do? Facts are facts. "There is no faith, however respectable, no interest, however legitimate, which must not accommodate itself to the progress of human knowledge and bend before truth" (in Count, 1950, p. 72). Paul Topinard, Broca's leading disciple and successor, took as his motto (1882, p. 748): *"J'ai horreur des systèmes et surtout des systèmes a priori"* (I abhor systems, especially a priori systems).

Broca singled out the few egalitarian scientists of his century for particularly harsh treatment because they had debased their calling by allowing an ethical hope or political dream to cloud their judgment and distort objective truth. "The intervention of political and social considerations has not been less injurious to anthropology than the religious element" (1855, in Count, 1950, p. 73). The great German anatomist Friedrich Tiedemann, for example, had argued that blacks and whites did not differ in cranial capacity. Broca nailed Tiedemann for the same error I uncovered in Morton's work (see pp. 50–69). When Morton used a subjective and imprecise method of reckoning, he calculated systematically lower capacities for blacks than when he measured the same skulls with a precise technique. Tiedemann, using an even more imprecise method, calculated a black average 45 cc above the mean value recorded by other scientists. Yet his measures for white skulls were no larger than those reported by colleagues. (For all his delight in exposing Tiedemann, Broca apparently never checked Morton's figures, though Morton was his hero and model. Broca once published a one-hundred-page paper analyzing Morton's techniques in the most minute detail—Broca, 1873b.)

Why had Tiedemann gone astray? "Unhappily," Broca wrote (1873b, p. 12), "he was dominated by a preconceived idea. He set out to prove that the cranial capacity of all human races is the same." But "it is an axiom of all observational sciences that facts must precede theories" (1868, p. 4). Broca believed, sincerely I assume, that facts were his only constraint and that his success in affirming traditional rankings arose from the precision of his measures and his care in establishing repeatable procedures.

Indeed, one cannot read Broca without gaining enormous respect for his care in generating data. I believe his numbers and doubt that any better have ever been obtained. Broca made an exhaustive study of all previous methods used to determine cranial capacity. He decided that lead shot, as advocated by "le célèbre Morton" (1861, p. 183), gave the best results, but he spent months refining the technique, taking into account such factors as the form and height of the cylinder used to receive the shot after it is poured from the skull, the speed of pouring shot into the skull, and the mode of shaking and tapping the skull to pack the shot and to determine whether or not more will fit in (Broca, 1873b). Broca finally developed an objective method for measuring cranial capacity. In most of his work, however, he preferred to weigh the brain directly after autopsies performed by his own hands.

I spent a month reading all of Broca's major work, concentrating on his statistical procedures. I found a definite pattern in his methods. He traversed the gap between fact and conclusion by what may be the usual route—predominantly in reverse. Conclusions came first and Broca's conclusions were the shared assumptions of most successful white males during his time—themselves on top by the good fortune of nature, and women, blacks, and poor people below. His facts were reliable (unlike Morton's), but they were gathered selectively and then manipulated unconsciously in the service of prior conclusions. By this route, the conclusions achieved not only the blessing of science, but the prestige of numbers. Broca and his school used facts as illustrations, not as constraining documents. They began with conclusions, peered through their facts, and came back in a circle to the same conclusions. Their example repays a closer study, for unlike Morton (who manipulated data, however unconsciously), they reflected their prejudices by another, and probably more common, route: advocacy masquerading as objectivity.

Selecting characters

When the "Hottentot Venus" died in Paris, Georges Cuvier, the greatest scientist and, as Broca would later discover to his delight, the largest brain of France, remembered this African woman as he had seen her in the flesh.

> She had a way of pouting her lips exactly like what we have observed in the orang-utan. Her movements had something abrupt and fantastical about them, reminding one of those of the ape. Her lips were monstrously large [those of apes are thin and small as Cuvier apparently forgot]. Her ear was like that of many apes, being small, the tragus weak, and the external border almost obliterated behind. These are animal characters. I have never seen a human head more like an ape than that of this woman (in Topinard, 1878, pp. 493–494).

The human body can be measured in a thousand ways. Any investigator, convinced beforehand of a group's inferiority, can select a small set of measures to illustrate its greater affinity with apes. (This procedure, of course, would work equally well for white males, though no one made the attempt. White people, for example, have thin lips—a property shared with chimpanzees—while most black Africans have thicker, consequently more "human," lips.)

Broca's cardinal bias lay in his assumption that human races could be ranked in a linear scale of mental worth. In enumerating the aims of ethnology, Broca included: "to determine the relative position of races in the human series" (in Topinard, 1878, p. 660). It did not occur to him that human variation might be ramified and random, rather than linear and hierarchical. And since he knew the order beforehand, anthropometry became a search for characters that would display the correct ranking, not a numerical exercise in raw empiricism.

Thus Broca began his search for "meaningful" characters—those that would display the established ranks. In 1862, for example, he tried the ratio of radius (lower arm bone) to humerus (upper arm bone), reasoning that a higher ratio marks a longer forearm—a character of apes. All began well: blacks yielded a ratio of .794, whites .739. But then Broca ran into trouble. An Eskimo skeleton yielded .703, an Australian aborigine .709, while the Hottentot Venus, Cuvier's near ape (her skeleton had been preserved in Paris), measured a mere .703. Broca now had two choices. He could either admit that, on this criterion, whites ranked lower than several dark-skinned groups, or he could abandon the criterion. Since he knew (1862a, p. 10) that Hottentots, Eskimos, and Australian aborigines ranked below most African blacks, he chose the second course: "After this, it seems difficult to me to continue to say that elongation of the forearm is a character of degradation or inferiority, because, on this account, the European occupies a place between Negroes on the one hand, and Hottentots, Australians, and eskimos on the other" (1862, p. 11).

Later, he almost abandoned his cardinal criterion of brain size because inferior yellow people scored so well:

> A table on which races were arranged by order of their cranial capacities would not represent the degrees of their superiority or inferiority, because size represents only one element of the problem [of ranking races]. On such a table, Eskimos, Lapps, Malays, Tartars and several other peoples of the Mongolian type would surpass the most civilized people of Europe. A lowly race may therefore have a big brain (1873a, p. 38).

But Broca felt that he could salvage much of value from his crude measure of overall brain size. It may fail at the upper end because some inferior groups have big brains, but it works at the lower end because small brains belong exclusively to people of low intelligence. Broca continued:

> But this does not destroy the value of small brain size as a mark of inferiority. The table shows that West African blacks have a cranial capacity about 100 cc less than that of European races. To this figure, we may add the following: Caffirs, Nubians, Tasmanians, Hottentots, Australians. These examples are sufficient to prove that if the volume of the brain does not play a decisive role in the intellectual ranking of races, it nevertheless has a very real importance (1873a, p. 38).

An unbeatable argument. Deny it at one end where conclusions are uncongenial; affirm it by the same criterion at the other. Broca did not fudge numbers; he merely selected among them or interpreted his way around them to favored conclusions.

In choosing among measures, Broca did not just drift passively in the sway of a preconceived idea. He advocated selection among characters as a stated goal with explicit criteria. Topinard, his chief disciple, distinguished between "empirical" characters "having no apparent design," and "rational" characters "related to some physiological opinion" (1878, p. 221). How then to determine which characters are "rational"? Topinard answered: "Other characteristics are looked upon, whether rightly or wrongly, as dominant.

They have an affinity in negroes to those which they exhibit in apes, and establish the transition between these and Europeans" (1878, p. 221). Broca had also considered this issue in the midst of his debate with Gratiolet, and had reached the same conclusion (1861, p. 176):

> We surmount the problem easily by choosing,for our comparison of brains, races whose intellectual inequalities are completely clear. Thus, the superiority of Europeans compared with African Negroes, American Indians, Hottentots, Australians and the Negroes of Oceania, is sufficiently certain to serve as a point of departure for the comparison of brains.

Particularly outrageous examples abound in the selection of individuals to represent groups in illustrations. Thirty years ago, when I was a child, the Hall of Man in the American Museum of Natural History still displayed the characters of human races by linear arrays running from apes to whites. Standard anatomical illustrations, until this generation, depicted a chimp, a Negro, and a white, part by part in that order—even though variation among whites and blacks is always large enough to generate a different order with other individuals: chimp, white, black. In 1903, for example, the American anatomist E. A. Spitzka published a long treatise on brain size and form in "men of eminence." He printed the following figure (Fig. 3.3) with a comment: "The jump from a Cuvier or a Thackeray to a Zulu or a Bushman is not greater than from the latter to the gorilla or the orang" (1903, p. 604). But he also published a similar figure (Fig. 3.4) illustrating variation in brain size among eminent whites apparently never realizing that he had destroyed his own argument. As F. P. Mall, the man who exposed Bean, wrote of these figures (1909, p. 24): "Comparing [them], it appears that Gambetta's brain resembles the gorilla's more than it does that of Gauss."

Averting anomalies

Inevitably, since Broca amassed so much disparate and honest data, he generated numerous anomalies and apparent exceptions to his guiding generality—that size of brain records intelligence and that comfortable white males have larger brains than women, poor people, and lower races. In noting how he worked around each apparent exception, we obtain our clearest insight into Broca's methods of argument and inference. We also understand why data could never overthrow his assumptions.

BIG-BRAINED GERMANS

Gratiolet, in his last desperate attempt, pulled out all the stops. He dared to claim that, on average, German brains are 100 grams heavier than French brains. Clearly, Gratiolet argued, brain size has nothing to do with intelligence! Broca responded disdainfully: "Monsieur Gratiolet has almost appealed to our patriotic sentiments. But it will be easy for me to show him that he can grant some value to the size of the brain without ceasing, for that, to be a good Frenchman" (1861, pp. 441–442).

Broca then worked his way systematically through the data. First of all, Gratiolet's figure of 100 grams came from unsupported claims of the German scientist E. Huschke. When Broca collated all the actual data he could find, the difference in size between German and French brains fell from 100 to 48 grams. Broca then applied a series of corrections for nonintellectual factors that also affect brain size. He argued, quite correctly, that brain size increases with body size, decreases with age, and decreases during long periods of poor health (thus explaining why executed criminals often have larger brains than honest folk who die of degenerative diseases in hospitals). Broca noted a mean French age of fifty-six and a half years in his sample, while the Germans averaged only fifty-one. He estimated that this difference would account for 16 grams of the disparity between French and Germans, cutting the German advantage to 32 grams. He then removed from the German sample all individuals who had died by violence or execution. The mean brain weight of twenty Germans, dead from natural causes, now stood at 1,320 grams, already *below* the French average of 1,333 grams. And Broca had not even yet corrected for the larger average body size of Germans. *Vive la France.*

Broca's colleague de Jouvencel, speaking on his behalf against the unfortunate Gratiolet, argued that greater German brawn accounted for all the apparent difference in brain and then some. Of the average German, he wrote (1861, p. 466):

> He ingests a quantity of solid food and drink far greater than that which satisfies us. This, joined with his consumption of beer, which is pervasive even in areas where wine is made, makes the German much more

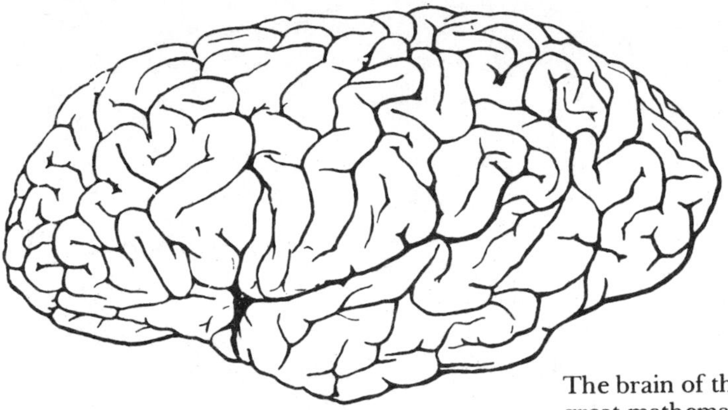

The brain of the
great mathematician
K. F. Gauss

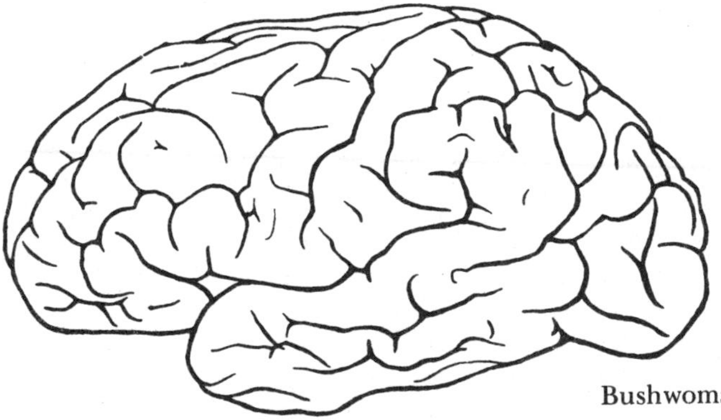

Bushwoman

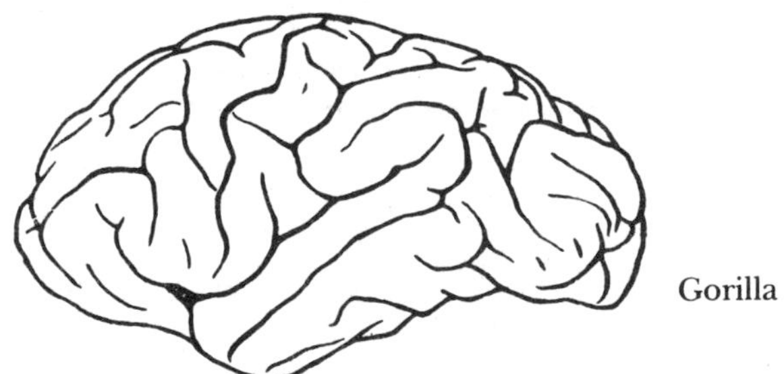

Gorilla

3 • 3 Spitzka's chain of being according to brain size.

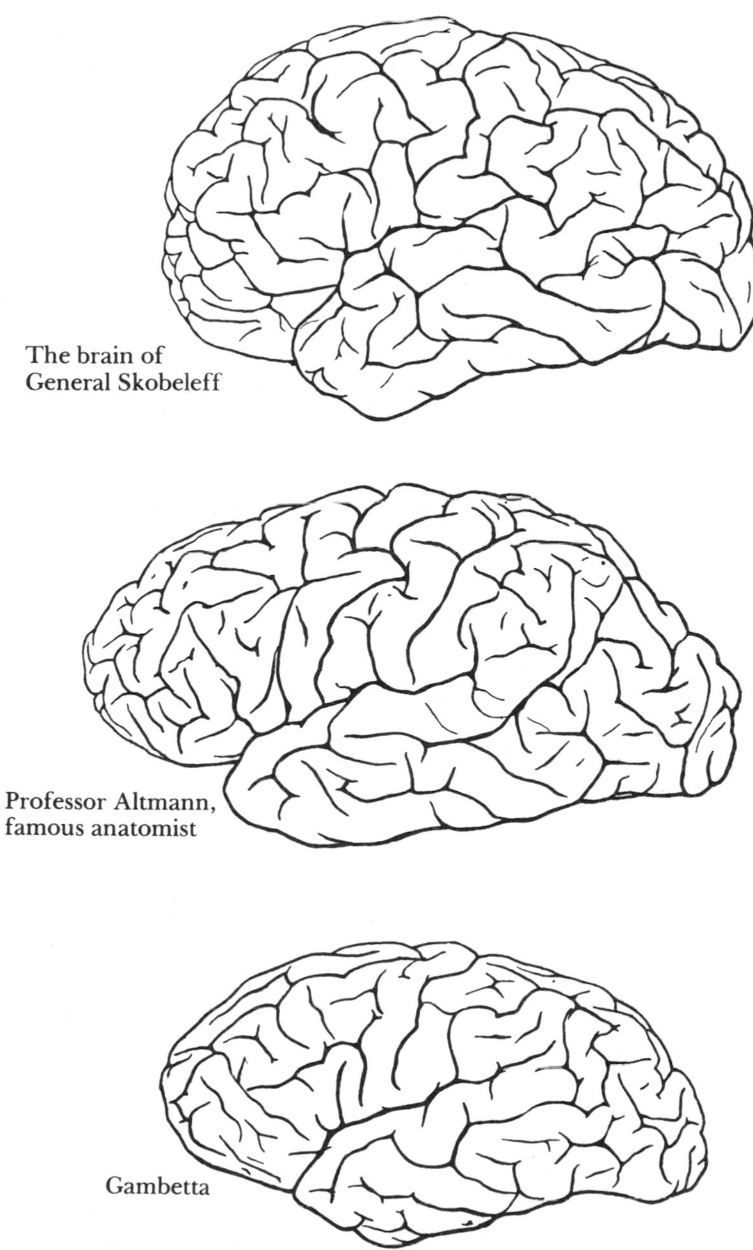

3 • 4 Spitzka's depiction of variation in brain size among white men of eminence.

fleshy [*charnu*] than the Frenchman—so much so that their relation of brain size to total mass, far from being superior to ours, appears to me, on the contrary, to be inferior.

I do not challenge Broca's use of corrections but I do note his skill in wielding them when his own position was threatened. Bear this in mind when I discuss how deftly he avoided them when they might have challenged a congenial conclusion—the small brains of women.

SMALL-BRAINED MEN OF EMINENCE

The American anatomist E. A. Spitzka urged men of eminence to donate their brains to science after their death. "To me the thought of an autopsy is certainly less repugnant than I imagine the process of cadaveric decomposition in the grave to be" (1907, p. 235). The dissection of dead colleagues became something of a cottage industry among nineteenth-century craniometricians. Brains exerted their customary fascination, and lists were proudly touted, accompanied by the usual invidious comparisons. (The leading American anthropologists J. W. Powell and W J McGee even made a wager over who carried the larger brain. As Ko-Ko told Nanki-Poo about the fireworks that would follow his execution, "You won't see them, but they'll be there all the same.")

Some men of genius did very well indeed. Against a European average of 1,300 to 1,400 grams, the great Cuvier stood out with his topheavy, 1,830 grams. Cuvier headed the charts until Turgenev finally broke the 2,000 gram barrier in 1883. (Other potential occupants of this stratosphere, Cromwell and Swift, lay in limbo for insufficiency of record.)

The other end was a bit more confusing and embarrassing. Walt Whitman managed to hear America singing with only 1,282 grams. As a crowning indignity, Franz Josef Gall, one of the two founders of phrenology—the original "science" of judging various mental capacities by the size of localized brain areas—weighed in at a meager 1,198 grams. (His colleague J. K. Spurzheim yielded a quite respectable 1,559 grams.) And, though Broca didn't know it, his own brain weighed only 1,424 grams, a bit above average to be sure, but nothing to crow about. Anatole France extended the range of famous authors to more than 1,000 grams when, in 1924, he opted for the other end of Turgenev's fame and clocked in at a mere 1,017 grams.

The small brains were troublesome, but Broca, undaunted, managed to account for all of them. Their possessors either died very old, were very short and slightly built, or had suffered poor preservation. Broca's reaction to a study by his German colleague Rudolf Wagner was typical. Wagner had obtained a real prize in 1855, the brain of the great mathematician Karl Friedrich Gauss. It weighed a modestly overaverage 1,492 grams, but was more richly convoluted than any brain previously dissected (Fig. 3.5). Encouraged, Wagner went on to weigh the brains of all dead and willing professors at Göttingen, in an attempt to plot the distribution of brain size among men of eminence. By the time Broca was battling with Gratiolet in 1861, Wagner had four more measurements. None posed any challenge to Cuvier, and two were distinctly puzzling—Hermann, the professor of philosophy at 1,368 grams, and Hausmann, the professor of mineralogy, at 1,226 grams. Broca corrected Hermann's brain for his age and raised it

3 • 5 The brain of the great mathematician K. F. Gauss (right) proved to be something of an embarrassment since, at 1,492 grams, it was only slightly larger than average. But other criteria came to the rescue. Here, E. A. Spitzka demonstrates that Gauss's brain is much more richly convoluted than that of a Papuan (left).

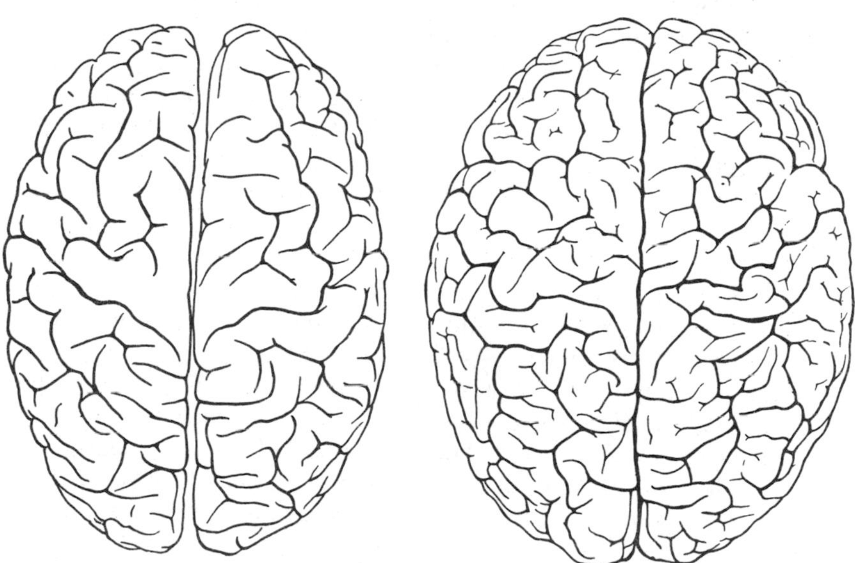

by 16 grams to 1.19 percent above average—"not much for a professor of linguistics," Broca admitted, "but still something" (1861, p. 167). No correction could raise Hausmann to the mean of ordinary folks, but considering his venerable seventy-seven years, Broca speculated that his brain may have undergone more than the usual amount of senile degeneration: "The degree of decadence that old age can impose upon a brain is very variable and cannot be calculated."

But Broca was still bothered. He could get around the low values, but he couldn't raise them to unusual weights. Consequently, to clinch an unbeatable conclusion, he suggested with a touch of irony that Wagner's post-Gaussian subjects may not have been so eminent after all:

> It is not very probable that 5 men of genius should have died within five years at the University of Göttingen. . . . A professorial robe is not necessarily a certificate of genius; there may be, even at Göttingen, some chairs occupied by not very remarkable men (1861, pp. 165–166).

At this point, Broca desisted: "The subject is delicate," he wrote (1861, p. 169), "and I must not insist upon it any longer."

LARGE-BRAINED CRIMINALS

The large size of many criminal brains was a constant source of bother to craniometricians and criminal anthropologists. Broca tended to dismiss it with his claim that sudden death by execution precluded the diminution that long bouts of disease produced in many honest men. In addition, death by hanging tended to engorge the brain and lead to spuriously high weights.

In the year of Broca's death, T. Bischoff published his study on the brains of 119 assassins, murderers, and thieves. Their average exceeded the mean of honest men by 11 grams, while 14 of them topped 1,500 grams, and 5 exceeded 1,600 grams. By contrast, only three men of genius could boast more than 1,600 grams, while the assassin Le Pelley, at 1,809 grams, must have given pause to the shade of Cuvier. The largest female brain ever weighed (1,565 grams) belonged to a woman who had killed her husband.

Broca's successor Paul Topinard puzzled over the data and finally decided that too much of a good thing is bad for some people. Truly inspired criminality may require as much upstairs as professorial virtuosity; who shall decide between Moriarty and Holmes? Topinard concluded: "It seems established that a certain proportion of criminals are pushed to depart from present social rules by an exuberance of cerebral activity and, consequently, by the fact of a large or heavy brain" (1888, p. 15).

FLAWS IN A PATTERN OF INCREASE THROUGH TIME

Of all Broca's studies, with the exception of his work on differences between men and women, none won more respect or attention than his supposed demonstration of steady increase in brain size as European civilization advanced from medieval to modern times (Broca, 1862b).

This study merits close analysis because it probably represents the best case of hope dictating conclusion that I have ever encountered. Broca viewed himself as a liberal in the sense that he did not condemn groups to permanent inferiority based on their current status. Women's brains had degenerated through time thanks to a socially enforced underusage; they might increase again under different social conditions. Primitive races had not been sufficiently challenged, while European brains grew steadily with the march of civilization.

Broca obtained large samples from each of three Parisian cemeteries, from the twelfth, the eighteenth, and the nineteenth centuries. Their average cranial capacities were, respectively, 1,426, 1,409, and 1,462 cc—not exactly the stuff for a firm conclusion of steady increase through time. (I have not been able to find Broca's raw data for statistical testing, but with a 3.5 percent mean difference between smallest and largest sample, it is likely that no statistically significant differences exist at all among the three samples.)

But how did these limited data—only three sites with no information on ranges of variation at a given time and no clear pattern through time—lead Broca to his hopeful conclusion? Broca himself admitted an initial disappointment: he had expected to find intermediate values in the eighteenth-century site (1862b, p. 106). Social class, he argued, must hold the answer, for successful groups within a culture owe at least part of their status to superior wits. The twelfth-century sample came from a churchyard and must represent gentry. A common grave provided the eighteenth-century skulls. But the nineteenth-century sample was a mixture, ninety skulls from individual graves with a mean of 1484 cc, and thirty-five from a common grave with an average of 1403 cc. Broca claimed that if differences in social class do not explain why calculated values fail to meet expectations, then the data are unintelligible. Intelligible, to Broca, meant steadily increasing through time—the proposition that the data were meant to prove, not rest upon. Again, Broca travels in a circle:

> Without this [difference in social class], we would have to believe that the cranial capacity of Parisians has really diminished during centuries following the 12th. Now during this period . . . intellectual and social progress has been considerable, and even if we are not yet certain that the development of civilization makes the brain grow as a consequence, no one, without doubt, would want to consider this cause as capable of making the brain decrease in size (1862b, p. 106).

But Broca's division of the nineteenth-century sample by social class also brought trouble as well as relief—for he now had two samples from common graves and the earlier one had a larger mean capacity, 1,409 for the eighteenth century vs. 1,403 for the nineteenth. But Broca was not to be defeated; he argued that the eighteenth-century common grave included a better class of people. In these prerevolutionary times, a man had to be really rich or noble to rest in a churchyard. The dregs of the poor measured 1,403 in the nineteenth century; the dregs leavened by good stock yielded about the same value one hundred years before.

Each solution brought Broca new trouble. Now that he was committed to a partition by social class within cemeteries, he had to admit that an additional seventeen skulls from the morgue's grave at the nineteenth-century site yielded a higher value than skulls of middle- and upper-class people from individual graves—1,517 vs. 1,484 cc. How could unclaimed bodies, abandoned to the state, surpass the cream of society? Broca reasoned in a chain of surpassingly weak inference: morgues stood on river borders; they probably housed a large number of drowned people; many drowned are suicides; many suicides are insane; many insane people, like criminals, have surprisingly large brains. With a bit of imagination, nothing can be truly anomalous.

Front and back

Tell me about this new young surgeon, Mr. Lydgate. I am told he is wonderfully clever; he certainly looks it—a fine brow indeed.

—GEORGE ELIOT, Middlemarch (1872)

Size of the whole, however useful and decisive in general terms, did not begin to exhaust the content of craniometry. Ever since the heyday of phrenology, specific parts of the brain and skull had been assigned definite status, thus providing a set of subsidiary criteria for the ranking of groups. (Broca, in his other career as a medical man, made his most important discovery in this area. In 1861 he developed the concept of cortical localization of function when he discovered that an aphasic patient had a lesion in the left inferior frontal gyrus, now called Broca's convolution.)

Most of these subsidiary criteria can be reduced to a single formula: front is better. Broca and his colleagues believed that higher mental functions were localized in anterior regions of the cortex, and that posterior areas busied themselves with the more mundane, though crucial, roles of involuntary movement, sensation, and emotion. Superior people should have more in front, less behind. We have already seen how Bean followed this assumption in generating his spurious data on front and back parts of the corpus callosum in whites and blacks.

Broca often used the distinction of front and back, particularly to extract himself from uncomfortable situations imposed by his data. He accepted Gratiolet's classification of human groups into *"races frontales"* (whites with anterior and frontal lobes most highly developed), *"races pariétales"* (Mongolians with parietal or mid lobes most prominent), and *"races occipitales"* (blacks with most in the back). He often unleashed the double whammy against inferior groups—small size and posterior prominence: "Negroes, and especially Hottentots, have a simpler brain than ours, and the relative poverty of their convolutions can be found primarily on their frontal lobes" (1873a, p. 32). As more direct evidence, he argued that Tahitians artificially deformed the frontal areas of certain male children in order to make the back portions bulge. These men became courageous warriors, but could never match white heroes for style: "Frontal deformation produced blind passions, ferocious instincts, and animal courage, all of which I would willingly call occipital courage. We must not confound it with true courage, frontal courage, which we may call Caucasian courage" (1861, pp. 202–203).

Broca also went beyond size to assess the quality of frontal vs. occipital regions in various races. Here, and not only to placate his adversary, he accepted Gratiolet's favorite argument that the sutures between skull bones close earlier in inferior races, thus trapping the brain within a rigid vault and limiting the effectiveness of further education. Not only do white sutures close later; they close in a different order—guess how? In blacks and other inferior people, the front sutures close first, the back sutures later; in whites, the front sutures close last. Extensive modern studies of cranial closure show no difference of timing or pattern among races (Todd and Lyon, 1924 and 1925).

Broca used this argument to extricate himself from a serious problem. He had described a sample of skulls from the earliest populations of *Homo sapiens* (Cro-Magnon type) and found that they exceeded modern Frenchmen in cranial capacity. Fortunately, however, their anterior sutures closed first and these progenitors must have been inferior after all: "These are signs of inferiority. We find them in all races in which the material life draws all cerebral activity to it. As intellectual life develops among a people, the anterior sutures become more complicated and stay open for a longer time" (1873a, p. 19).

The argument of front and back,* so flexible and far-ranging, served as a powerful tool for rationalizing prejudice in the face of apparently contradictory fact. Consider the following two examples.

THE CRANIAL INDEX

Beyond brain size itself, the two most hoary and misused measures of craniometry were surely the facial angle (jutting forward of face and jaws—the less the better), and the cranial index. The cranial index never had much going for it beyond ease of measurement. It was calculated as the ratio of maximum width to maximum

*Broca did not confine his arguments on the relative worth of brain parts to the distinction between front and back. Virtually any measured difference between peoples could be given a value in terms of prior conviction about relative worth. Broca once claimed, for example (1861, p. 187), that blacks probably had larger cranial nerves than whites, hence a larger nonintellectual portion of the brain.

length of the skull. Relatively long skulls (ratio of .75 or less) were called dolichocephalic; relatively short skulls (over .8), brachycephalic. Anders Retzius, the Swedish scientist who popularized the cranial index, constructed a theory of civilization upon it. He believed that Stone Age peoples of Europe were brachycephalic, and that progressive Bronze Age elements (Indo-European, or Aryan dolichocephalics) later invaded and replaced the original and more primitive inhabitants. Some original brachycephalic stocks survive among such benighted people as Basques, Finns, and Lapps.

Broca disproved this popular tale conclusively by discovering dolichocephalics both among Stone Age skulls and within modern remnants of "primitive" stocks. Indeed, Broca had good reason to be suspicious of attempts by Nordic and Teutonic scientists to enshrine dolichocephaly as a mark of higher capability. Most Frenchmen, including Broca himself (Manouvrier, 1899), were brachycephalic. In a passage that recalls his dismissal of Tiedemann's claims for equality between black and white brains, Broca labeled Retzius's doctrine as self-serving gratification rather than empirical truth. Did he ever consider the possibility that he might fall prey to similar motivations?

> Since the work of Mr. Retzius, scientists have generally held, without sufficient study, that dolichocephaly is a mark of superiority. Perhaps so; but we must also not forget that the characters of dolichocephaly and brachycephaly were studied first in Sweden, then in England, the United States and Germany—and that in all these countries, particularly in Sweden, the dolichocephalic type clearly predominates. It is a natural tendency of men, even among those most free of prejudice, to attach an idea of superiority to the dominant characteristics of their race (1861, p. 513).

Obviously, Broca declined to equate brachycephaly with inherent stupidity. Still, the prestige of dolichocephaly was so great that Broca felt more than a little uncomfortable when clearly inferior people turned up longheaded—uncomfortable enough to invent one of his most striking, unbeatable arguments. The cranial index had run into a stunning difficulty: not only were African blacks and Australian aborigines dolichocephalic, but they turned out to be the world's most longheaded peoples. Adding insult to this injury, the fossil Cro-Magnon skulls were not only larger than those of modern Frenchmen; they were more dolichocephalic as well.

Dolichocephaly, Broca reasoned, could be attained in several ways. The longheadedness that served as a mark of Teutonic genius obviously arose by frontal elongation. Dolichocephalics among people known to be inferior must have evolved by lengthening the back—occipital dolichocephaly in Broca's terms. With one sweep, Broca encompassed both the superior cranial capacity and the dolichocephaly of his Cro-Magnon fossils: "It is by the greater development of their posterior cranium that their general cranial capacity is rendered greater than ours" (1873a, p. 41). As for blacks, they had acquired both a posterior elongation and a diminution in frontal width, thus giving them both a smaller brain in general and a longheadedness (not to be confused with the Teutonic style) exceeded by no human group. As to the brachycephaly of Frenchmen, it is no failure of frontal elongation (as the Teutonic supremacists claimed), but an addition of width to a skull already admirable.

THE CASE OF THE FORAMEN MAGNUM

The foramen magnum is the hole in the base of our skull. The spinal cord passes through it and the vertebral column articulates to the bone around its edge (the occipital condyle). In the embryology of all mammals, the foramen magnum begins under the skull, but migrates back to a position behind the skull at birth. In humans, the foramen magnum migrates only slightly and remains under the skull in adults. The foramen magnum of adult great apes occupies an intermediate position, not so far forward as in humans, not so far back as in other mammals. The functional significance of these orientations is clear. An upright animal like *Homo sapiens* must have its skull mounted *on top* of its vertebral column in order to look forward when standing erect; fourfooted animals mount their vertebral column *behind* their skull and look forward in their usual posture.

These differences provided an irresistible source for invidious comparison. Inferior peoples should have a more posterior foramen magnum, as in apes and lower mammals. In 1862 Broca entered an existing squabble on this issue. Relative egalitarians like James Cowles Pritchard had been arguing that the foramen magnum lies exactly in the center of the skull in both whites and blacks. Racists like J. Virey had discovered graded variation, the higher the race, the more forward the foramen magnum. Neither side, Broca noted, had much in the way of data. With characteristic objectivity, he set out to resolve this vexatious, if minor, issue.

Broca amassed a sample of sixty whites and thirty-five blacks and measured the length of their skulls both before and behind the anterior border of the foramen magnum. Both races had the same amount of skull behind—100.385 mm for whites, 100.857 mm for blacks (note precision to third decimal place). But whites had much less in front (90.736 vs. 100.304 mm) and their foramen magnum therefore lay in a more anterior position (see Table 3.1). Broca concluded: "In orang-utans, the posterior projection [the part of the skull behind the foramen magnum] is shorter. It is therefore incontestable . . . that the conformation of the Negro, in this respect as in many others, tends to approach that of the monkey" (1862c, p. 16).

But Broca then began to worry. The standard argument about the foramen magnum referred only to its relative position on the cranium itself, not to the face projecting in front of the cranium. Yet Broca had included the face in his anterior measure. Now everyone knows, he wrote, that blacks have longer faces than whites. This is an apelike sign of inferiority in its own right, but it should not be confused with the relative position of the foramen magnum within the cranium. Thus Broca set out to subtract the facial influence from his measures. He found that blacks did, indeed, have longer faces—white faces accounted for only 12.385 mm of their anterior measure, black faces for 27.676 mm (see Table 3.1). Subtracting facial length, Broca obtained the following figures for anterior cranium: 78.351 for whites, 72.628 for blacks. In other words, based on the cranium alone, the foramen magnum

Table 3 • 1 *Broca's measurements on the relative position of the foramen magnum*

	WHITES	BLACKS	DIFFERENCE IN FAVOR OF BLACKS
ANTERIOR	90.736	100.304	+ 9.568
Facial	12.385	27.676	+15.291
Cranial	78.351	72.628	− 5.723
POSTERIOR	100.385	100.857	+ 0.472

of blacks lay *farther forward* (the ratio of front to back, calculated from Broca's data, is .781 for whites, and .720 for blacks). Clearly, by criteria explicitly accepted before the study, blacks are superior to whites. Or so it must be, unless the criteria suddenly shift, as they did forthwith.

The venerable argument of front and back appeared to rescue Broca and the threatened people he represented. The more forward position of the foramen magnum in blacks does not record their superiority after all; it only reflects their lack of anterior brain power. Relative to whites, blacks have lost a great deal of brain in front. But they have added some brain behind, thus reducing the front/back ratio of the foramen magnum and providing a spurious appearance of black advantage. But they have not added to these inferior back regions as much as they lost in the anterior realm. Thus blacks have smaller and more poorly proportioned brains than whites:

> The anterior cranial projection of whites . . . surpasses that of Negroes by 4.9 percent. . . . Thus, while the foramen magnum of Negroes is further back with respect to their incisors [Broca's most forward point in his anterior measure that included the face], it is, on the contrary, further forward with respect to the anterior edge of their brain. To change the cranium of a white into that of a Negro, we would have not only to move the jaws forward, but also to reduce the front of the cranium—that is, to make the anterior brain atrophy and to give, as insufficient compensation, part of the material we extracted to the posterior cranium. In other words, in Negroes, the facial and occipital regions are developed to the detriment of the frontal region (1862c, p. 18).

This was a small incident in Broca's career, but I can imagine no better illustration of his method—shifting criteria to work through good data toward desired conclusions. Heads I'm superior; tails, you're inferior.

And old arguments never seem to die. Walter Freeman, dean of American lobotomists (he performed or supervised thirty-five hundred lesions of frontal portions of the brain before his retirement in 1970), admitted late in his career (cited in Chorover, 1979):

> What the investigator misses most in the more highly intelligent individuals is their ability to introspect, to speculate, to philosophize, especially in regard to onesself. . . . On the whole, psychosurgery reduces creativity, sometimes to the vanishing point.

Freeman then added that "women respond better than men, Negroes better than whites." In other words, people who didn't have as much up front in the first place, don't miss it as badly.

Women's brains

Of all his comparisons between groups, Broca collected most information on the brains of women vs. men—presumably because it was more accessible, not because he held any special animus toward women. "Inferior" groups are interchangeable in the general theory of biological determinism. They are continually juxtaposed, and one is made to serve as a surrogate for all—for the general proposition holds that society follows nature, and that social rank reflects innate worth. Thus, E. Huschke, a German anthropologist, wrote in 1854: "The Negro brain possesses a spinal cord of the type found in children and women and, beyond this, approaches the type of brain found in higher apes" (in Mall, 1909, pp. 1–2). The celebrated German anatomist Carl Vogt wrote in 1864:

> By its rounded apex and less developed posterior lobe the Negro brain resembles that of our children, and by the protuberance of the parietal lobe, that of our females. . . . The grown-up Negro partakes, as regards his intellectual faculties, of the nature of the child, the female, and the senile white. . . . Some tribes have founded states, possessing a peculiar organization; but, as to the rest, we may boldly assert that the whole race has, neither in the past nor in the present, performed anything tending to the progress of humanity or worthy of preservation (1864, pp. 183–192).

G. Hervé, a colleague of Broca, wrote in 1881: "Men of the black races have a brain scarcely heavier than that of white women" (1881, p. 692). I do not regard as empty rhetoric a claim that the battles of one group are for all of us.

Broca centered his argument about the biological status of modern women upon two sets of data: the larger brains of men in modern societies and a supposed widening through time of the disparity in size between male and female brains. He based his most extensive study upon autopsies he performed in four Parisian hospitals. For 292 male brains, he calculated a mean weight of 1,325 grams; 140 female brains averaged 1,144 grams for a difference of 181 grams, or 14 percent of the male weight. Broca understood, of course, that part of this difference must be attributed to the larger size of males. He had used such a correction to rescue Frenchmen from a claim of German superiority (p. 89). In that case, he knew how to make the correction in exquisite detail. But now he made no attempt to measure the effect of size alone, and actually stated that he didn't need to do so. Size, after all, cannot account for the entire difference because we know that women are not as intelligent as men.

We might ask if the small size of the female brain depends exclusively upon the small size of her body. Tiedemann has proposed this explanation. But we must not forget that women are, on the average, a little less intelligent than men, a difference which we should not exaggerate but which is, nonetheless, real. We are therefore permitted to suppose that the relatively small size of the female brain depends in part upon her physical inferiority and in part upon her intellectual inferiority (1861, p. 153).

To record the supposed widening of the gap through time, Broca measured the cranial capacities of prehistoric skulls from L'Homme Mort cave. Here he found a difference of only 99.5 cc between males and females, while modern populations range from 129.5 to 220.7 cc. Topinard, Broca's chief disciple, explained the increasing discrepancy through time as a result of differing evolutionary pressures upon dominant men and passive women:

The man who fights for two or more in the struggle for existence, who has all the responsibility and the cares of tomorrow, who is constantly active in combatting the environment and human rivals, needs more brain than the woman whom he must protect and nourish, than the sedentary woman, lacking any interior occupations, whose role is to raise children, love, and be passive (1888, p. 22).

In 1879 Gustave Le Bon, chief misogynist of Broca's school, used these data to publish what must be the most vicious attack upon women in modern scientific literature (it will take some doing to beat Aristotle). Le Bon was no marginal hate-monger. He was a founder of social psychology and wrote a study of crowd behavior still cited and respected today (*La psychologie des foules,* 1895). His writings also had a strong influence upon Mussolini. Le Bon concluded:

In the most intelligent races, as among the Parisians, there are a large number of women whose brains are closer in size to those of gorillas than to the most developed male brains. This inferiority is so obvious that no one can contest it for a moment; only its degree is worth discussion. All psychologists who have studied the intelligence of women, as well as poets and novelists, recognize today that they represent the most inferior forms of human evolution and that they are closer to children and savages than to an adult, civilized man. They excel in fickleness, inconstancy, absence of thought and logic, and incapacity to reason. Without doubt there exist some distinguished women, very superior to the average man, but they are as exceptional as the birth of any monstrosity, as, for example, of a gorilla with two heads; consequently, we may neglect them entirely (1879, pp. 60–61).

Nor did Le Bon shrink from the social implications of his views. He was horrified by the proposal of some American reformers to grant women higher education on the same basis as men:

A desire to give them the same education, and, as a consequence, to propose the same goals for them, is a dangerous chimera. . . . The day when, misunderstanding the inferior occupations which nature has given her, women leave the home and take part in our battles; on this day a social revolution will begin, and everything that maintains the sacred ties of the family will disappear (1879, p. 62).

Sound familiar?*

I have reexamined Broca's data, the basis for all this derivative pronouncement, and I find the numbers sound but Broca's interpretation, to say the least, ill founded. The claim for increasing difference through time is easily dismissed. Broca based this contention on the sample from L'Homme Mort alone. It consists of seven male, and six female, skulls. Never has so much been coaxed from so little!

In 1888 Topinard published Broca's more extensive data on Parisian hospitals. Since Broca recorded height and age as well as brain size, we may use modern statistical procedures to remove their effect. Brain weight decreases with age, and Broca's women were, on average, considerably older than his men at death. Brain weight increases with height, and his average man was almost half a foot taller than his average woman. I used multiple regression, a technique that permits simultaneous assessment of the influence of

*Ten years later, America's leading evolutionary biologist, E. D. Cope, dreaded the result if "a spirit of revolt become general among women." "Should the nation have an attack of this kind," he wrote (1890, p. 2071), "like a disease, it would leave its traces in many after-generations." He detected the beginnings of such anarchy in pressures exerted by women "to prevent men from drinking wine and smoking tobacco in moderation," and in the carriage of misguided men who supported female suffrage: "Some of these men are effeminate and long-haired."

height and age upon brain size. In an analysis of the data for women, I found that, at average male height and age, a woman's brain would weigh 1,212 grams.* Correction for height and age reduces the 181 gram difference by more than a third to 113 grams.

It is difficult to assess this remaining difference because Broca's data contain no information about other factors known to influence brain size in a major way. Cause of death has an important effect, as degenerative disease often entails a substantial diminution of brain size. Eugene Schreider (1966), also working with Broca's data, found that men killed in accidents had brains weighing, on average, 60 grams more than men dying of infectious diseases. The best modern data that I can find (from American hospitals) records a full 100 gram difference between death by degenerative heart disease and by accident or violence. Since so many of Broca's subjects were elderly women, we may assume that lengthy degenerative disease was more common among them than among the men.

More importantly, modern students of brain size have still not agreed on a proper measure to eliminate the powerful effect of body size (Jerison, 1973; Gould, 1975). Height is partly adequate, but men and women of the same height do not share the same body build. Weight is even worse than height, because most of its variation reflects nutrition rather than intrinsic size—and fat vs. skinny exerts little influence upon the brain. Léonce Manouvrier took up this subject in the 1880s and argued that muscular mass and force should be used. He tried to measure this elusive property in various ways and found a marked difference in favor of men, even in men and women of the same height. When he corrected for what he called "sexual mass," women came out slightly ahead in brain size.

Thus, the corrected 113 gram difference is surely too large; the true figure is probably close to zero and may as well favor women as men. One hundred thirteen grams, by the way, is exactly the average difference between a five-foot four-inch and a six-foot-four-inch male in Broca's data†—and we would not want to ascribe

*I calculate, where y is brain size in grams, x_1 age in years, and x_2 body height in cm: $y = 764.5 - 2.55x_1 + 3.47x_2$

† For his largest sample of males, and using the favored power function for bivariate

greater intelligence to tall men. In short, Broca's data do not permit any confident claim that men have bigger brains than women.

Maria Montessori did not confine her activities to educational reform for young children. She lectured on anthropology for several years at the University of Rome and wrote an influential book entitled *Pedagogical Anthropology* (English edition, 1913). She was, to say the least, no egalitarian. She supported most of Broca's work and the theory of innate criminality proposed by her compatriot Cesare Lombroso (next chapter). She measured the circumference of children's heads in her schools and inferred that the best prospects had bigger brains. But she had no use for Broca's conclusions about women. She discussed Manouvrier's work at length and made much of his tentative claim that women have slightly larger brains when proper corrections are made. Women, she concluded, are intellectually superior to men, but men have prevailed heretofore by dint of physical force. Since technology has abolished force as an instrument of power, the era of women may soon be upon us: "In such an epoch there will really be superior human beings, there will really be men strong in morality and in sentiment. Perhaps in this way the reign of woman is approaching, when the enigma of her anthropological superiority will be deciphered. Woman was always the custodian of human sentiment, morality and honor" (1913, p. 259).

Montessori's argument represents one possible antidote to "scientific" claims for the constitutional inferiority of certain groups. One may affirm the validity of biological distinctions, but argue that the data have been misinterpreted by prejudiced men with a stake in the outcome, and that disadvantaged groups are truly superior. In recent years, Elaine Morgan has followed this strategy in her *Descent of Woman,* a speculative reconstruction of human prehistory from the woman's point of view—and as farcical as more famous tall tales by and for men.

I dedicate this book to a different position. Montessori and Morgan follow Broca's method to reach a more congenial conclusion. I would rather label the whole enterprise of setting a biological value upon groups for what it is: irrelevant, intellectually unsound, and highly injurious.

---

analysis of brain allometry, I calculate, where y is brain weight in grams and x is body height in cm: $y = 121.6x^{0.47}$

Postscript

Craniometric arguments lost much of their luster in our century, as determinists switched their allegiance to intelligence testing—a more "direct" path to the same invalid goal of ranking groups by mental worth—and as scientists exposed the prejudiced nonsense that dominated most literature on form and size of the head. The American anthropologist Franz Boas, for example, made short work of the fabled cranial index by showing that it varied widely both among adults of a single group and within the life of an individual (Boas, 1899). Moreover, he found significant differences in cranial index between immigrant parents and their American-born children. The immutable obtuseness of the brachycephalic southern European might veer toward the dolichocephalic Nordic norm in a single generation of altered environment (Boas, 1911).

Yet the supposed intellectual advantage of bigger heads refuses to disappear entirely as an argument for assessing human worth. We still encounter it occasionally at all levels of determinist contention.

1. Variation within the general population: Arthur Jensen (1979, pp. 361–362) supports the value of IQ as a measure of innate intelligence by claiming that the correlation between brain size and IQ is about 0.30. He doesn't doubt that the correlation is meaningful and that "there has been a direct causal effect, through natural selection in the course of human evolution, between intelligence and brain size." Undaunted by the low value of the correlation, he proclaims that it would be even higher if so much of the brain were not "devoted to noncognitive functions."

On the same page, Jensen cites an average correlation of 0.25 between IQ and physical stature. Although this value is effectively the same as the IQ vs. brain size correlation, Jensen switches ground and holds that "this correlation almost certainly involves no causal or functional relationship between stature and intelligence." Both height and intelligence, he argues, are perceived as desirable traits, and people lucky enough to possess more than the average of both are drawn to each other. But is it not more likely that height vs. brain size represents the primary causal correlation for the obvious reason that tall people tend to have large body parts? Brain size would then be an imperfect measure of height, and IQ might correlate with it (at the low value of 0.3) for the primarily environmental reason that poverty and poor nutrition can lead both to reduced stature and poor IQ scores.

2. Variation among social classes and occupational groups: In a book dedicated to putting educators in touch with latest advances in the brain sciences, H. T. Epstein (in Chall and Mirsky, 1978) states (pp. 349–350):

> First we shall ask if there is any indication of a linkage of any kind between brain and intelligence. It is generally stated that there is no such linkage. . . . But the one set of data I have found seems to show clearly that there is a substantial connection. Hooton studied the head circumferences of white Bostonians as part of his massive study of criminals. The following table shows that the ordering of people according to head size yields an entirely plausible ordering according to vocational status. It is not at all clear how the impression has been spread that there is no such correlation.

Epstein's chart, reproduced as he presents it in Table 3.2, seems to support the notion that people in more prestigious jobs have larger heads. But a bit of probing and checking in original sources exposes the chart as a shoddy bit of finagling (not by Epstein who, I suspect, copied it from another secondary source that I have not been able to identify).

i) Epstein's reported standard deviations are so low, and therefore imply such a small range of variation within each occupational class, that the differences in mean head size must be significant

Table 3 • 2 *Mean and standard deviation of head circumference for people of varied vocational statuses*

VOCATIONAL STATUS	N	MEAN (IN MM)	S.D.
Professional	25	569.9	1.9
Semiprofessional	61	566.5	1.5
Clerical	107	566.2	1.1
Trades	194	565.7	0.8
Public service	25	564.1	2.5
Skilled trades	351	562.9	0.6
Personal services	262	562.7	0.7
Laborers	647	560.7	0.3

Source: Ernest A. Hooton, *The American Criminal*, vol. 1 (Cambridge, Mass.: Harvard University Press, 1939), Table VIII-17.

even though they are so small. But a glance at Hooton's original table (1939, Table VIII-17) reveals that the wrong column (standard errors of the mean) has been copied and called standard deviation. The true standard deviations, given in another column of Hooton's table, run from 14.4 to 18.6—large enough to render most mean differences between occupational groups statistically insignificant.

ii) The chart arranges occupational groups by mean head size, but does not include Hooton's ranked assessments of vocational status based upon years of education (1939, p. 150). In fact, since the column is labeled "vocational status," we are led to assume that the jobs have been listed in their proper order of prestige and that a perfect correlation therefore exists between status and head size. But the professions are arranged only by head size. Several professions do not fit the pattern; personal services and skilled trades (Hooton's status 5 and 6) rank just above the bottom in head size but at the middle in prestige.

iii) As a much worse, and entirely inexcusable omission, my consultation of Hooton's original chart shows that data for three trades have been expunged without comment in Table 3.2. Guess why? All three rank at or near the bottom of Hooton's list of status—factory workers at rank 7 (of 11), transportation employees at rank 8, and "extractive" trades (farming and mining) at the lowest rank 11. All three have mean head circumferences (564.7, 564.9, and 564.7, respectively) *above* the grand average for all professions (563.9)!

I do not know the source of this disgracefully fudged chart. Jensen (1979, p. 361) reproduces it in Epstein's version with the three trades omitted. But he correctly labels the standard error (though he also omits the standard deviation) and properly denotes the professions as "occupational category" rather than "vocational status." Yet Jensen's version includes the same minor numerical error as Epstein's (standard error of 0.3 for laborers, miscopied as the correct value from the omitted line of "extractive" workers placed just above laborers in Hooton's chart). Since I doubt that the same insignificant error would have been made twice independently, and since Jensen's book and Epstein's article appeared at virtually the same time, I assume that both took the information from an unidentified secondary source (neither cite anyone but Hooton).

iv) Since Epstein and Jensen make so much of Hooton's data, they might have consulted his own opinion about it. Hooton was no do-gooding environmentalist liberal. He was a strong eugenicist and biological determinist who ended his study of American criminals with these chilling words: "The elimination of crime can be effected only by the extirpation of the physically, mentally, and morally unfit, or by their complete segregation in a socially aseptic environment" (1939, p. 309). Yet Hooton himself thought that his chart of head sizes and professions had proved nothing (1939, p. 154). He noted that only one vocational group, laborers, departed significantly from the average of all groups. And he stated explicitly that his sample for the only profession with noticeably larger than average heads—the professionals—was "wholly inadequate" (p. 153) as a result of its small size.

v) The primary environmental hypothesis for correlations of head size with social class holds that they are artifacts of a causal correlation between body size and status. Large bodies tend to carry large heads, and proper nutrition and freedom from poverty fosters better growth in childhood. Hooton's data provide tentative support for both parts of this argument, though Epstein doesn't mention these data on stature at all. Hooton provides information on both height and weight (both inadequate measures of stature—see p. 106). Most significant deviations from the grand average support the environmental hypothesis. For weight, two groups departed significantly: professionals (status 1) heavier than average, and laborers (status 10) lighter than average. For height, three groups were deficient and none significantly taller than average: laborers (status 10), personal service (status 5), and clerical (status 2—and contrary to the environmentalist hypothesis). I also computed correlation coefficients for head circumference vs. stature from Hooton's data. I found no correlation for total height, but significant correlations for both sitting height (0.605) and weight (0.741).

3. Variation among races: In its eighteenth edition of 1964, the *Encyclopaedia Britannica* was still listing "a small brain in relation to their size" along with woolly hair as characteristic of black people.

In 1970 the South African anthropologist P. V. Tobias wrote a courageous article exposing the myth that group differences in brain size bear any relationship to intelligence—indeed, he argued, group differences in brain size, independent of body size and other biasing factors, have never been demonstrated at all.

This conclusion may strike readers as strange, especially since it comes from a famous scientist well acquainted with the reams of published data on brain size. After all, what can be simpler than weighing a brain?—Take it out, and put it on the scale. Not so. Tobias lists fourteen important biasing factors. One set refers to problems of measurement itself: at what level is the brain severed from the spinal cord; are the meninges removed or not (meninges are the brain's covering membranes, and the dura mater, or thick outer covering, weighs 50 to 60 grams); how much time elapsed after death; was the brain preserved in any fluid before weighing and, if so, for how long; at what temperature was the brain preserved after death. Most literature does not specify these factors adequately, and studies made by different scientists usually cannot be compared. Even when we can be sure that the same object has been measured in the same way under the same conditions, a second set of biases intervenes—influences upon brain size with no direct tie to the desired properties of intelligence or racial affiliation: sex, body size, age, nutrition, nonnutritional environment, occupation, and cause of death. Thus, despite thousands of published pages, and tens of thousands of subjects, Tobias concludes that we do not know—as if it mattered at all—whether blacks, on the average, have larger or smaller brains than whites. Yet the larger size of white brains was an unquestioned "fact" among white scientists until quite recently.

Many investigators have devoted an extraordinary amount of attention to the subject of group differences in human brain size. They have gotten nowhere, not because there are no answers, but because the answers are so difficult to get and because the a priori convictions are so clear and controlling. In the heat of Broca's debate with Gratiolet, one of Broca's defenders, admittedly as a nasty debating point, made a remark that admirably epitomizes the motivations implicit in the entire craniometric tradition: "I have noticed for a long time," stated de Jouvencel (1861, p. 465), "that, in general, those who deny the intellectual importance of the brain's volume have small heads." Self-interest, for whatever reason, has been the wellspring of opinion on this heady issue from the start.

FOUR

Measuring Bodies

Two Case Studies on the Apishness of Undesirables

The concept of evolution transformed human thought during the nineteenth century. Nearly every question in the life sciences was reformulated in its light. No idea was ever more widely used, or misused ("social Darwinism" as an evolutionary rationale for the inevitability of poverty, for example). Both creationists (Agassiz and Morton) and evolutionists (Broca and Galton) could exploit the data of brain size to make their invalid and invidious distinctions among groups. But other quantitative arguments arose as more direct spinoffs from evolutionary theory. In this chapter I discuss two as representatives of a prevalent type; they present both a strong contrast and an interesting similarity. The first is the most general evolutionary defense of all for ranking groups—the argument from recapitulation, often epitomized by the obfuscating tongue-twister "ontogeny recapitulates phylogeny." The second is a specific evolutionary hypothesis for the biological nature of human criminal behavior—Lombroso's criminal anthropology. Both theories relied upon the same quantitative and supposedly evolutionary method—the search for signs of apish morphology in groups deemed undesirable.

The ape in all of us: recapitulation

Once the fact of evolution had been established, nineteenth-century naturalists devoted themselves to tracing the actual pathways that evolution had followed. They sought, in other words, to reconstruct the tree of life. Fossils might have provided the evidence, for only they could record the actual ancestors of modern forms. But the fossil record is extremely imperfect, and the major trunks and branches of life's tree all grew before the evolution of hard parts permitted the preservation of a fossil record at all. Some indirect criterion had to be found. Ernst Haeckel, the great German zoologist, refurbished an old theory of creationist biology and suggested that the tree of life might be read directly from the embryological development of higher forms. He proclaimed that "ontogeny recapitulates phylogeny" or, to explicate this mellifluous tongue-twister, that an individual, in its own growth, passes through a series of stages representing *adult* ancestral forms in their correct order—an individual, in short, climbs its own family tree.

Recapitulation ranks among the most influential ideas of late nineteenth-century science. It dominated the work of several professions, including embryology, comparative morphology, and paleontology. All these disciplines were obsessed with the idea of reconstructing evolutionary lineages, and all regarded recapitulation as the key to this quest. The gill slits of an early human embryo represented an ancestral adult fish; at a later stage, the temporary tail revealed a reptilian or mammalian ancestor.

Recapitulation spilled forth from biology to influence several other disciplines in crucial ways. Both Sigmund Freud and C. G. Jung were convinced recapitulationists, and Haeckel's idea played no small role in the development of psychoanalytic theory. (In *Totem and Taboo,* for example, Freud tries to reconstruct human history from a central clue provided by the Oedipus complex of young boys. Freud reasoned that this urge to parricide must reflect an actual event among ancestral adults. Hence, the sons of an ancestral clan must once have killed their father in order to gain access to women.) Many primary-school curriculums of the late nineteenth century were reconstructed in the light of recapitulation. Several school boards prescribed the *Song of Hiawatha* in early grades, reasoning that children, passing through the savage stage of their ancestral past, would identify with it.*

*Readers interested in the justification provided for recapitulation by Haeckel and his colleagues, and in the reasons for its later downfall, may consult my dull, but highly detailed treatise, *Ontogeny and Phylogeny,* Harvard University Press, 1977.

Recapitulation also provided an irresistible criterion for any scientist who wanted to rank human groups as higher and lower. The *adults* of *inferior* groups must be like *children* of *superior* groups, for the child represents a primitive adult ancestor. If adult blacks and women are like white male children, then they are living representatives of an ancestral stage in the evolution of white males. An anatomical theory for ranking races—based on entire bodies, not only on heads—had been found.

Recapitulation served as a general theory of biological determinism. All "inferior" groups—races, sexes, and classes—were compared with the children of white males. E. D. Cope, the celebrated American paleontologist who elucidated the mechanism of recapitulation (see Gould, 1977, pp. 85–91), identified four groups of lower human forms on this criterion: nonwhite races, all women, southern as opposed to northern European whites, and lower classes within superior races (1887, pp. 291–293—Cope particularly despised "the lower classes of the Irish"). Cope preached the doctrine of Nordic supremacy and agitated to curtail the immigration of Jews and southern Europeans to America. To explain the inferiority of southern Europeans in recapitulatory terms, he argued that warmer climates impose an earlier maturation. Since maturation signals the slowdown and cessation of bodily development, southern Europeans are caught in a more childlike, hence primitive, state as adults. Superior northerners move on to higher stages before a later maturation cuts off their development:

> There can be little doubt that in the Indo-European race maturity in some respects appears earlier in tropical than in northern regions; and though subject to many exceptions, this is sufficiently general to be looked upon as a rule. Accordingly, we find in that race—at least in the warmer regions of Europe and America—a larger proportion of certain qualities which are more universal in women, as greater activity of the emotional nature when compared with the judgment. . . . Perhaps the more northern type left all that behind in its youth (1887, pp. 162–163).

Recapitulation provided a primary focus for anthropometric, particularly craniometric, arguments about the ranking of races. The brain, once again, played a dominant role. Louis Agassiz, in a creationist context, had already compared the brain of adult blacks with that of a white fetus seven months old. We have already cited

(p. 103) Vogt's remarkable statement equating the brains of adult blacks and white women with those of white male children and explaining, on this basis, the failure of black people to build any civilization worthy of his notice.

Cope also focused upon the skull, particularly upon "those important elements of beauty, a well-developed nose and beard" (1887, pp. 288–290), but he also derided the deficient calf musculature of blacks:

> Two of the most prominent characters of the negro are those of immature stages of the Indo-European race in its characteristic types. The deficient calf is the character of infants at a very early stage; but, what is more important, the flattened bridge of the nose and shortened nasal cartilages are universally immature conditions of the same parts in the Indo-European. . . . In some races—e.g., the Slavic—this undeveloped character persists later than in some others. The Greek nose, with its elevated bridge, coincides not only with aesthetic beauty, but with developmental perfection.

In 1890 American anthropologist D. G. Brinton summarized the argument with a paean of praise for measurement:

> The adult who retains the more numerous fetal, infantile or simian traits, is unquestionably inferior to him whose development has progressed beyond them. . . . Measured by these criteria, the European or white race stands at the head of the list, the African or negro at its foot. . . . All parts of the body have been minutely scanned, measured and weighed, in order to erect a science of the comparative anatomy of the races (1890, p. 48).

If anatomy built the hard argument of recapitulation, psychic development offered a rich field for corroboration. Didn't everyone know that savages and women are emotionally like children? Despised groups had been compared with children before, but the theory of recapitulation gave this old chestnut the respectability of main-line scientific theory. "They're like children" was no longer just a metaphor of bigotry; it now embodied a theoretical claim that inferior people were literally mired in an ancestral stage of superior groups.

G. Stanley Hall, then America's leading psychologist, stated the general argument in 1904: "Most savages in most respects are children, or, because of sexual maturity, more properly, adolescents of adult size" (1904, vol. 2, p. 649). A. F. Chamberlain, his chief disciple, opted for the paternalistic mode: "Without primitive peoples, the world at large would be much what in small it is without the blessing of children."

The recapitulationists extended their argument to an astonishing array of human capacities. Cope compared prehistoric art with the sketches of children and living "primitives" (1887, p. 153): "We find that the efforts of the earliest races of which we have any knowledge were quite similar to those which the untaught hand of infancy traces on its slate or the savage depicts on the rocky faces of cliffs." James Sully, a leading English psychologist, compared the aesthetic senses of children and savages, but gave the edge to children (1895, p. 386):

> In much of this first crude utterance of the aesthetic sense of the child we have points of contact with the first manifestations of taste in the race. Delight in bright, glistening things, in gay things, in strong contrasts of color, as well as in certain forms of movement, as that of feathers—the favorite personal adornment—this is known to be characteristic of the savage and gives to his taste in the eyes of civilized man the look of childishness. On the other hand, it is doubtful whether the savage attains to the sentiment of the child for the beauty of flowers.

Herbert Spencer, the apostle of social Darwinism, offered a pithy summary (1895, pp. 89–90): "The intellectual traits of the uncivilized . . . are traits recurring in the children of the civilized."

Since recapitulation became a focus for the general theory of biological determinism, many male scientists extended the argument to women. E. D. Cope claimed that the "metaphysical characteristics" of women were

> . . . very similar in essential nature to those which men exhibit at an early stage of development. . . . The gentler sex is characterized by a greater impressibility; . . . warmth of emotion, submission to its influence rather than that of logic; timidity and irregularity of action in the outer world. All these qualities belong to the male sex, as a general rule, at some period of life, though different individuals lose them at very various periods. . . . Probably most men can recollect some early period of their lives when the emotional nature predominated—a time when emotion at the sight of suffering was more easily stirred than in maturer years. . . . Perhaps all men can recall a period of youth when they were hero-worshippers—when they felt the need of a stronger arm, and loved to look up to the powerful

friend who could sympathize with and aid them. This is the "woman stage" of character (1887, p. 159).

In what must be the most absurd statement in the annals of biological determinism, G. Stanley Hall—again, I remind you, not a crackpot, but America's premier psychologist—invoked the higher suicide rates of women as a sign of their primitive evolutionary status (1904, vol. 2, p. 194):

> This is one expression of a profound psychic difference between the sexes. Woman's body and soul is phyletically older and more primitive, while man is more modern, variable, and less conservative. Women are always inclined to preserve old customs and ways of thinking. Women prefer passive methods; to give themselves up to the power of elemental forces, as gravity, when they throw themselves from heights or take poison, in which methods of suicide they surpass man. Havelock Ellis thinks drowning is becoming more frequent, and that therein women are becoming more womanly.

As a justification for imperialism, recapitulation offered too much promise to remain sequestered in academic pronouncements. I have already cited Carl Vogt's low opinion of African blacks, based on his comparison of their brains with those of white children. B. Kidd extended the argument to justify colonial expansion into tropical Africa (1898, p. 51). We are, he wrote, "dealing with peoples who represent the same stage in the history of the development of the race that the child does in the history of the development of the individual. The tropics will not, therefore, be developed by the natives themselves."

In the course of a debate about our right to annex the Philippines, Rev. Josiah Strong, a leading American imperialist, piously declared that "our policy should be determined not by national ambition, nor by commercial considerations, but by our duty to the world in general and to the Filipinos in particular" (1900, p. 287). His opponents, citing Henry Clay's contention that the Lord would not create a people incapable of self-government, argued against the need for our benevolent tutelage. But Clay had spoken in the bad old days before evolutionary theory and recapitulation:

> Clay's conception was formed . . . before modern science had shown that races develop in the course of centuries as individuals do in years, and that an undeveloped race, which is incapable of self-government, is no

more of a reflection on the Almighty than is an undeveloped child who is incapable of self-government. The opinions of men who in this enlightened day believe that the Filipinos are capable of self-government because everybody is, are not worth considering.

Even Rudyard Kipling, the poet laureate of imperialism, used the recapitulationist argument in the first stanza of his most famous apology for white supremacy:

Take up the White Man's Burden

Send forth the best ye breed
Go, bind your sons to exile
to serve the captive's need:
To wait, in heavy harness,
On fluttered folk and wild—
Your new-caught sullen peoples,
Half devil and half child.

Teddy Roosevelt, whose judgment was not always so keen, wrote to Henry Cabot Lodge that the verse "was very poor poetry but made good sense from the expansion point of view" (in Weston, 1972, p. 35).

And so the story might stand, a testimony to nineteenth-century folly and prejudice, if an interesting twist had not been added during our own century. By 1920 the theory of recapitulation had collapsed (Gould, 1977, pp. 167–206). Not long after, the Dutch anatomist Louis Bolk proposed a theory of exactly opposite meaning. Recapitulation required that adult traits of ancestors develop more rapidly in descendants to become juvenile features—hence, traits of modern children are primitive characters of ancestral adults. But suppose that the reverse process occurs as it often does in evolution. Suppose that juvenile traits of ancestors develop so slowly in descendants that they become adult features. This phenomenon of retarded development is common in nature; it is called neoteny (literally, "holding on to youth"). Bolk argued that humans are essentially neotenous. He listed an impressive set of features shared by adult humans and fetal or juvenile apes, but lost in adult apes: vaulted cranium and large brain in relation to body size; small face; hair confined largely to head, armpits, and pubic regions; unrotated big toe. I have already discussed one of the most important signs of human neoteny in another context (pp. 101–

103): retention of the foramen magnum in its fetal position, under the skull.

Now consider the implications of neoteny for the ranking of human groups. Under recapitulation, adults of inferior races are like children of superior races. But neoteny reverses the argument. In the context of neoteny, it is "good"—that is, advanced or superior—to retain the traits of childhood, to develop more slowly. Thus, superior groups retain their childlike characters as adults, while inferior groups pass through the higher phase of childhood and then degenerate toward apishness. Now consider the conventional prejudice of white scientists: whites are superior, blacks inferior. Under recapitulation, black adults should be like white children. But under neoteny, white adults should be like black children.

For seventy years, under the sway of recapitulation, scientists had collected reams of objective data all loudly proclaiming the same message: adult blacks, women, and lower-class whites are like white upper-class male children. With neoteny now in vogue, these hard data could mean only one thing: upper-class adult males are inferior because they lose, while other groups retain, the superior traits of childhood. There is no escaping it.

At least one scientist, Havelock Ellis, did bow to the clear implication and admit the superiority of women, though he wriggled out of a similar confession for blacks. He even compared rural with urban men, found that men of the city were developing womanly anatomy, and proclaimed the superiority of urban life (1894, p. 519): "The large-headed, delicate-faced, small-boned man of urban civilization is much nearer to the typical woman than is the savage. Not only by his large brain, but by his large pelvis, the modern man is following a path first marked out by woman." But Ellis was iconoclastic and controversial (he wrote one of the first systematic studies of sexuality), and his application of neoteny to sexual differences never made much impact. Meanwhile, with respect to racial differences, supporters of human neoteny adopted another, more common, tactic: they simply abandoned their seventy years of hard data and sought new and opposite information to confirm the inferiority of blacks.

Louis Bolk, chief defender of human neoteny, declared that the most strongly neotenized races are superior. In retaining more juvenile features, they have kept further away from "the pithecoid ancestor of man" (1929, p. 26). "From this point of view, the division of mankind into higher and lower races is fully justified [1929, p. 26]. It is obvious that I am, on the basis of my theory, a convinced believer in the inequality of races" (1926, p. 38). Bolk reached into his anatomical grab-bag and extracted some traits indicating a greater departure for black adults from the advantageous proportions of childhood. Led by these new facts to an old and comfortable conclusion, Bolk proclaimed (1929, p. 25): "The white race appears to be the most progressive, as being the most retarded." Bolk, who viewed himself as a "liberal" man, declined to relegate blacks to permanent ineptitude. He hoped that evolution would be benevolent to them in the future:

> It is possible for all other races to reach the zenith of development now occupied by the white race. The only thing required is continued progressive action in these races of the biological principle of anthropogenesis [i.e., neoteny]. In his fetal development the negro passes through a stage that has already become the final stage for the white man. Well then, when retardation continues in the negro too, what is still a transitional stage may for this race also become a final one (1926, pp. 473–474).

Bolk's argument verged on the dishonest for two reasons. First, he conveniently forgot all the features—like the Grecian nose and full beard so admired by Cope—that recapitulationists had stoutly emphasized because they placed whites *far* from the conditions of childhood. Secondly, he sidestepped a pressing and embarrassing issue: Orientals, not whites, are clearly the most neotenous of human races (Bolk listed the neotenous features of both races selectively and then proclaimed the differences too close to call; see Ashley Montagu, 1962, for a fairer assessment). Women, moreover, are more neotenous than men. I trust that I will not be seen as vulgar white apologist if I decline to press the superiority of Oriental women and declare instead that the whole enterprise of ranking groups by degree of neoteny is fundamentally unjustified. Just as Anatole France and Walt Whitman could write as well as Turgenev with brains about half the weight of his, I would be more than mildly surprised if the small differences in degree of neoteny among races bear any relationship to mental ability or moral worth.

Nonetheless, old arguments never die. In 1971 the British psychologist and genetic determinist H. J. Eysenck again brought forth a neotenic argument for black inferiority. Eysenck took three facts and used neoteny to forge a story from them: 1) black babies and young children exhibit more rapid sensorimotor development than whites—that is, they are less neotenic because they depart more quickly from the fetal state; 2) average white IQ surpasses average black IQ by age three; 3) there is a slight negative correlation between sensorimotor development in the first year of life and later IQ—that is, children who develop more rapidly tend to end up with lower IQ's. Eysenck concludes (1971, p. 79): "These findings are important because of a very general view in biology [the theory of neoteny] according to which the more prolonged the infancy the greater in general are the cognitive or intellectual abilities of the species. This law appears to work even within a given species."

Eysenck fails to realize that he has based his argument on what is almost surely a noncausal correlation. (Noncausal correlations are the bane of statistical inference—see Chapter 6. They are perfectly "true" in a mathematical sense, but they demonstrate no causal connection. For example, we may calculate a spectacular correlation—very near the maximum value of 1.0—between the rise in world population during the past five years and the increasing separation of Europe and North America by continental drift.) Suppose that lower black IQ is purely a result of generally poorer environment. Rapid sensorimotor development is one way of identifying a person as black—but a less accurate way than skin color. The correlation of poor environment with lower IQ may be causal, but the correlation of rapid sensorimotor development with lower IQ is probably noncausal because rapid sensorimotor development, in this context, merely identifies a person as black. Eysenck's argument ignores the fact that black children, in a racist society, generally live in poorer environments, which may lead to lower IQ scores. Yet Eysenck invoked neoteny to give theoretical meaning, and thereby causal status, to a noncausal correlation reflecting his hereditarian bias.

The ape in some of us: criminal anthropology

Atavism and criminality

In *Resurrection,* Tolstoy's last great novel (1899), the assistant prosecutor, an unfeeling modernist, rises to condemn a prostitute falsely accused of murder:

> The assistant prosecutor spoke at great length. . . . All the latest catch-phrases then in vogue in his set, everything that then was and still is accepted as the last word in scientific wisdom was included in his speech—heredity and congenital criminality, Lombroso and Tarde, evolution and the struggle for existence. . . . "Running away with himself, isn't he?" said the presiding judge with a smile, bending towards the austere member of the court. "A fearful dunderhead!" said the austere member.

In Bram Stoker's *Dracula* (1897), Professor Van Helsing urges Mina Harker to describe the evil Count: "Tell us . . . dry men of science what you see with those so bright eyes." She responds: "The Count is a criminal and of criminal type. Nordau and Lombroso would so classify him, and qua criminal he is of imperfectly formed mind."*

Maria Montessori expressed an embattled optimism when she wrote in 1913 (p. 8): "The phenomenon of criminality spreads without check or succor, and up to yesterday it aroused in us nothing but repulsion and loathing. But now that science has laid its finger upon this moral fester, it demands the cooperation of all mankind to combat it."

The common subject of these disparate assessments is Cesare Lombroso's theory of *l'uomo delinquente*—the criminal man—probably the most influential doctrine ever to emerge from the anthropometric tradition. Lombroso, an Italian physician, described the insight that led to his theory of innate criminality and to the profession he established—criminal anthropology. He had, in 1870, been trying to discover anatomical differences between criminals and insane men "without succeeding very well." Then, "the morning of a gloomy day in December," he examined the skull of the famous

* In his *Annotated Dracula*, Leonard Wolf (1975, p. 300) notes that Jonathan Harker's initial description of Count Dracula is based directly upon Cesare Lombroso's account of the born criminal. Wolf presents the following contrasts:

HARKER WRITES: "His [the Count's] face was . . . aquiline, with high bridge of the thin nose and peculiarly arched nostrils. . . ."

LOMBROSO: "[The criminal's] nose on the contrary . . . is often aquiline like the beak of a bird of prey."

HARKER: "His eyebrows were very massive, almost meeting over the nose. . . ."

LOMBROSO: "The eyebrows are bushy and tend to meet across the nose."

HARKER: ". . . his ears were pale and at the tops extremely pointed. . . ."

LOMBROSO: "with a protuberance on the upper part of the posterior margin . . . a relic of the pointed ear. . . ."

brigand Vihella, and had that flash of joyous insight that marks both brilliant discovery and crackpot invention. For he saw in that skull a series of atavistic features recalling an apish past rather than a human present:

> This was not merely an idea, but a flash of inspiration. At the sight of that skull, I seemed to see all of a sudden, lighted up as a vast plain under a flaming sky, the problem of the nature of the criminal—an atavistic being who reproduces in his person the ferocious instincts of primitive humanity and the inferior animals. Thus were explained anatomically the enormous jaws, high cheek bones, prominent superciliary arches, solitary lines in the palms, extreme size of the orbits, handle-shaped ears found in criminals, savages and apes, insensibility to pain, extremely acute sight, tattooing, excessive idleness, love of orgies, and the irresponsible craving of evil for its own sake, the desire not only to extinguish life in the victim, but to mutilate the corpse, tear its flesh and drink its blood (in Taylor et al., 1973, p. 41).

Lombroso's theory was not just a vague proclamation that crime is hereditary—such claims were common enough in his time—but a specific *evolutionary* theory based upon anthropometric data. Criminals are evolutionary throwbacks in our midst. Germs of an ancestral past lie dormant in our heredity. In some unfortunate individuals, the past comes to life again. These people are innately driven to act as a normal ape or savage would, but such behavior is deemed criminal in our civilized society. Fortunately, we may identify born criminals because they bear anatomical signs of their apishness. Their atavism is both physical and mental, but the physical signs, or stigmata as Lombroso called them, are decisive. Criminal *behavior* can also arise in normal men, but we know the "born criminal" by his anatomy. Anatomy, indeed, is destiny, and born criminals cannot escape their inherited taint: "We are governed by silent laws which never cease to operate and which rule society with more authority than the laws inscribed on our statute books. Crime . . . appears to be a natural phenomenon" (Lombroso, 1887, p. 667).

Animals and savages as born criminals

The identification of apish atavism in criminals did not clinch Lombroso's argument, for physical apishness can explain a man's barbaric behavior only if the natural inclinations of savages and lower animals are criminal. If some men look like apes, but apes be kind, then the argument fails. Thus, Lombroso devoted the first part of his major work (*Criminal Man,* first published in 1876) to what must be the most ludicrous excursion into anthropomorphism ever published—an analysis of the criminal behavior of animals. He cites, for example, an ant driven by rage to kill and dismember an aphid; an adulterous stork who, with her lover, murdered her husband; a criminal association of beavers who ganged up to murder a solitary compatriot; a male ant, without access to female reproductives, who violated a (female) worker with atrophied sexual organs, causing her great pain and death; he even refers to the insect eating of certain plants as an "equivalent of crime" (Lombroso, 1887, pp. 1–18).

Lombroso then proceeded to the next logical step: comparison of criminals with "inferior" groups. "I would compare," wrote a French supporter, "the criminal to a savage appearing, by atavism, in modern society; we may think that he was born a criminal because he was born a savage" (Bordier, 1879, p. 284). Lombroso ventured into ethnology to identify criminality as normal behavior among inferior people. He wrote a small treatise (Lombroso, 1896) on the Dinka of the Upper Nile. In it, he spoke of their heavy tattooing and low threshold for pain—at puberty they break their incisors with a hammer. They display apish stigmata as normal parts of their anatomy: "their nose . . . is not only flattened, but trilobed, resembling that of monkeys." His colleague G. Tarde wrote that some criminals "would have been the ornament and the moral aristocracy of a tribe of Red Indians" (in Ellis, 1910, p. 254). Havelock Ellis made much of a claim that criminals and inferior people often do not blush. "Inability to blush has always been considered the accompaniment of crime and shamelessness. Blushing is also very rare among idiots and savages. The Spaniards used to say of the South American Indians: 'How can one trust men who do not know how to blush' " (1910, p. 138). And how far did the Incas get by trusting Pizarro?

Lombroso constructed virtually all his arguments in a manner that precluded their defeat, thus making them scientifically vacuous. He cited copious numerical data to lend an air of objectivity to his work, but it remained so vulnerable that even most of Broca's school turned against the theory of atavism. Whenever Lombroso encountered a contrary fact, he performed some mental gymnastics to incorporate it within his system. This posture is clearly expressed in his statements on the depravity of inferior peoples, for again and again he encountered stories of courage and accomplishment among those he wished to denigrate. Yet he twisted all these stories into his system. If, for example, he had to admit a favorable trait, he joined it with others he could despise. Citing the somewhat dated authority of Tacitus for his conclusion, he wrote: "Even when honor, chastity, and pity are found among savages, impulsiveness and laziness are never wanting. Savages have a horror of continuous work, so that for them the passage to active and methodical labor lies by the road of selection or of slavery only" (1911, p. 367). Or consider his one begrudging word of praise for the inferior and criminal race of gypsies:

> They are vain, like all delinquents, but they have no fear or shame. Everything they earn they spend for drink and ornaments. They may be seen barefooted, but with bright-colored or lace-bedecked clothing; without stockings, but with yellow shoes. They have the improvidence of the savage and that of the criminal as well. . . . They devour half-putrified carrion. They are given to orgies, love a noise, and make a great outcry in the markets. They murder in cold blood in order to rob, and were formerly suspected of cannibalism. . . . It is to be noted that this race, so low morally and so incapable of cultural and intellectual development, a race that can never carry on any industry, and which in poetry has not got beyond the poorest lyrics, has created in Hungary a marvelous musical art—a new proof of the genius that, mixed with atavism, is to be found in the criminal (1911, p. 40).

If he had no damning traits to mix with his praise, he simply discounted the motivation for apparently worthy behavior among "primitives." A white saint dying bravely under torture is a hero among heroes; a "savage" expiring with equal dignity simply doesn't feel the pain:

> Their [criminals'] physical insensibility well recalls that of savage peoples who can bear in rites of puberty, tortures that a white man could never endure. All travellers know the indifference of Negroes and American savages to pain: the former cut their hands and laugh in order to avoid work; the latter, tied to the torture post, gaily sing the praises of their tribe while they are slowly burnt (1887, p. 319).

We recognize in this comparison of atavistic criminals with animals, savages, and people of lower races the basic argument of recapitulation discussed in the previous section. To complete the chain, Lombroso needed only to proclaim the child as inherently criminal—for the child is an ancestral adult, a living primitive. Lombroso did not shrink from this necessary implication, and he branded as criminal the traditional innocent of literature: "One of the most important discoveries of my school is that in the child up to a certain age are manifested the saddest tendencies of the criminal man. The germs of delinquency and of criminality are found normally even in the first periods of human life" (1895, p. 53). Our impression of the child's innocence is a class bias; we comfortable folks suppress the natural inclinations of our children: "One who lives among the upper classes has no idea of the passion babies have for alcoholic liquor, but among the lower classes it is only too common a thing to see even suckling babes drink wine and liquors with wonderful delight" (1895, p. 56).*

The stigmata: anatomical, physiological, and social

Lombroso's anatomical stigmata (Fig. 4.1) were, for the most part, neither pathologies nor discontinuous variations, but extreme values on a normal curve that approach average measures for the same trait in great apes. (In modern terms, this is a fundamental source of Lombroso's error. Arm length varies among humans, and some people must have longer arms than others. The average chimp has a longer arm than the average human, but this doesn't mean that a relatively long-armed human is genetically similar to apes. Normal variation *within* a population is a different biological phenomenon from differences in average values *between* populations. This error occurs again and again. It is the basis of Arthur Jensen's fallacy in asserting that average differences in IQ between American whites and blacks are largely inherited—see

*In *Dracula,* Professor Van Helsing, in his inimitable broken English, extolled the argument from recapitulation by branding the Count as a persistent child (and therefore both a primitive and a criminal as well):

> Ah! there I have hope that our man-brains that have been of man so long and that have not lost the grace of God, will come higher than his child-brain that lie in his tomb for centuries, that grow not yet to our stature, and that do only work selfish and therefore small. . . . He is clever and cunning and resourceful; but he be not of man-stature as to brain. He be of child-brain in much. Now this criminal of ours is predestinate to crime also; he too have child-brain, and it is of the child to do what he have done. The little bird, the little fish, the little animal learn not by principle but empirically; and when he learn to do, then there is to him the ground to start from to do more.

A
B
C
center
D
E
F
G

pp. 156–157. A true atavism is a discontinuous, genetically based, ancestral trait—the occasional horse born with functional side toes, for example.) Among his apish stigmata, Lombroso listed (1887, pp. 660–661): greater skull thickness, simplicity of cranial sutures, large jaws, preeminence of the face over the cranium, relatively long arms, precocious wrinkles, low and narrow forehead, large ears, absence of baldness, darker skin, greater visual acuity, diminished sensitivity to pain, and absence of vascular reaction (blushing). At the 1886 International Congress on Criminal Anthropology, he even argued (see Fig. 4.2) that the feet of prostitutes are often prehensile as in apes (big toe widely separated from others).

For other stigmata, Lombroso descended from the apes to seek

4•2 The feet of prostitutes. This figure was presented by L. Jullien to the 4th International Congress on Criminal Anthropology in 1896. Commenting upon it, Lombroso said: "These observations show admirably that the morphology of the prostitute is more abnormal even than that of the criminal, especially for atavistic anomalies, because the prehensile foot is atavistic."

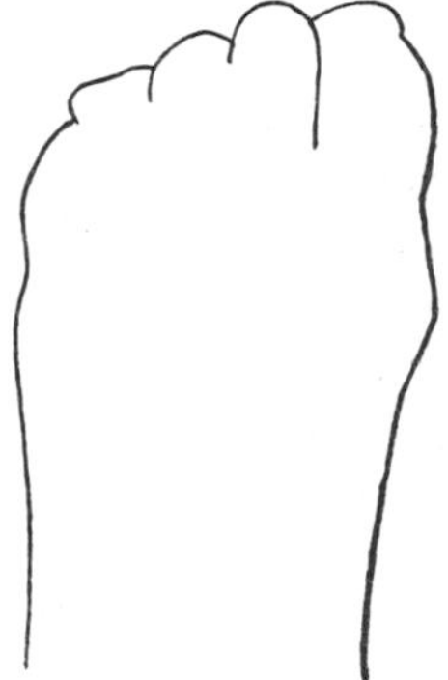

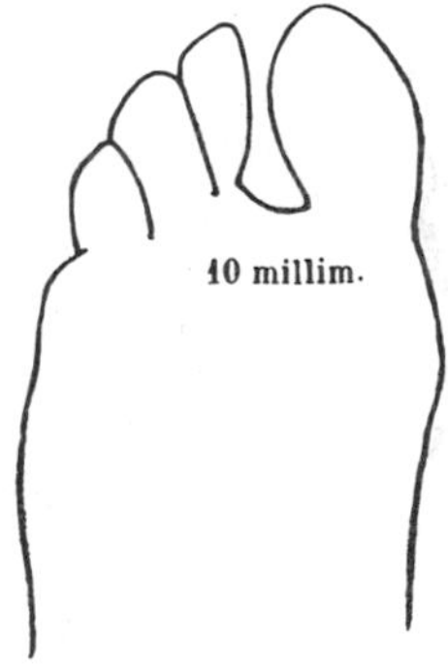

4•1 A panoply of criminal faces. The frontispiece to the atlas of Lombroso's *Criminal Man.* Group E are German murderers; Group I are burglars (Lombroso tells us that the man without a nose managed to escape justice for many years by wearing the false nose depicted in the figure on the left, wearing a derby); "H" are purse snatchers; "A" are shoplifters; "B," "C," "D," and "F" are swindlers; while the distinguished gentlemen of the bottom row declared themselves bankrupt fraudulently.

similarity with more distant, and even more "primitive," creatures: he compared prominent canine teeth and a flattened palate with the anatomy of lemurs and rodents, an oddly shaped occipital condyle (area for articulation of skull and vertebral column) with the normal condyles of cattle and pigs (1896, p. 188), and an abnormal heart with the usual conformation in sirenians (a group of rare marine mammals). He even postulated a meaningful similarity between the facial asymmetry of some criminals and flatfishes with both eyes on the upper surface of their bodies (1911, p. 373)!

Lombroso bolstered his study of specific defects with a general anthropometric survey of the criminal head and body—a sample of 383 crania from dead criminals, plus general proportions measured for 3,839 among the living. As an indication of Lombroso's style, consider the numerical basis of his most important claim—that criminals generally have smaller brains than normal people, even though a few criminals have very large brains (see p. 94).* Lombroso (1911, p. 365) and his disciples (Ferri, 1897, p. 8, for example) repeated this claim continually. Yet Lombroso's data show no such thing. Fig. 4.3 presents the frequency distributions for cranial capacity measured by Lombroso in 121 male criminals and 328 upright men. You don't need fancy statistics to see that the two distributions differ very little—despite Lombroso's conclusion that, in criminals, "the small capacities dominate and the very great are rare" (1887, p. 144). I have reconstructed the original data from Lombroso's tables of percentages within classes and calculate average values of 1,450 cc for criminal heads and 1,484 cc for law-abiding heads. The standard deviations of the two distributions (a general measure of spread about the average) do not differ significantly. This means that the larger range of variation in the law-abiding sample—an important point for Lombroso since it extended the maximum capacity for decent folk to 100 cc above

*Other standard craniometrical arguments were often pressed into service by criminal anthropology. For example, as early as 1843, Voisin invoked the classical argument of front and back (see pp. 97–103) to place criminals among the animals. He studied five hundred young offenders and reported deficiencies in the forward and upper parts of their brain—the supposed seat of morality and rationality. He wrote (1843, pp. 100–101):

> Their brains are at a minimum of development in their anterior and superior parts, in the two parts that make us what we are and that place us above the animals and make us men. They [criminal brains] are placed by their nature . . . entirely outside the human species.

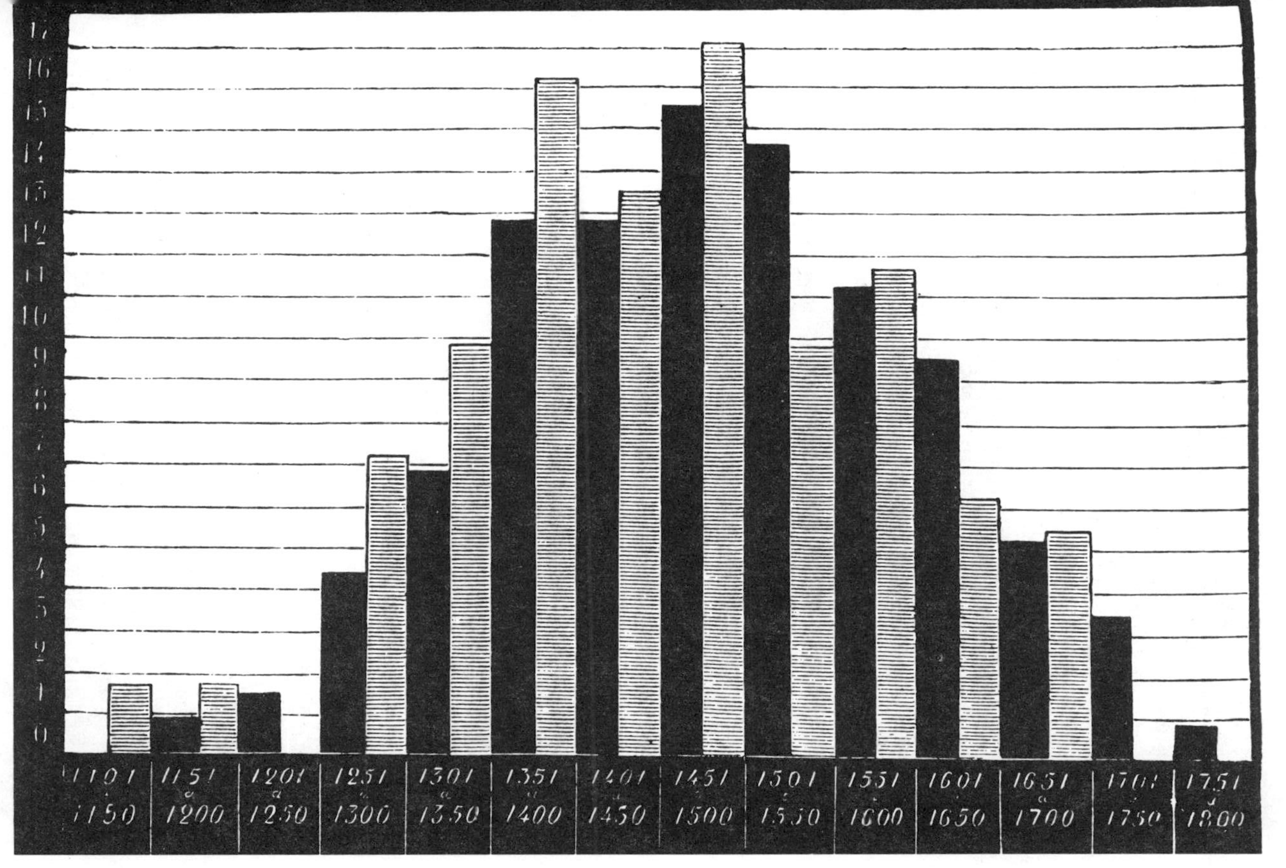

4 • 3 The cranial capacities of normal men (in black) compared with criminals (hatched). The y-axis is in percentages rather than actual numbers.

the maximum for criminals—may simply be an artifact of larger sample size for law-abiding men (the larger the sample, the greater the chance of including extreme values).

Lombroso's stigmata also included a set of social traits. He emphasized particularly: 1) The argot of criminals, a language of their own with high levels of onomatopoeia, much like the speech of children and savages: "Atavism contributes to it more than anything else. They speak differently because they feel differently; they speak as savages because they are true savages in the midst of our brilliant European civilization" (1887, p. 476); 2) Tattooing, reflecting both the insensitivity of criminals to pain and their atavistic love of adornment (Fig. 4.4). Lombroso made a quantitative study of content in criminal tattoos and found them, in general, lawless ("vengeance") or excusing ("born under an unlucky star," "out of luck"), though he encountered one that read: "Long live France and french fried potatoes."

Lombroso never attributed all criminal acts to people with atavistic stigmata. He concluded that about 40 percent of criminals followed hereditary compulsion; others acted from passion, rage, or desperation. At first glance, this distinction of occasional from born criminals has the appearance of a compromise or retreat, but Lombroso used it in an opposite way—as a claim that rendered his system immune to disproof. No longer could men be characterized by their acts. Murder might be a deed of the lowest ape in a human body or of the most upright cuckold overcome by justified rage. All criminal acts are covered: a man with stigmata performs them by innate nature, a man without stigmata by force of circumstances. By classifying exceptions within his system, Lombroso excluded all potential falsification.

Lombroso's retreat

Lombroso's theory of atavism caused a great stir and aroused one of the most heated scientific debates of the nineteenth century. Lombroso, though he peppered his work with volumes of numbers, had not made the usual obeisances to cold objectivity. Even those great a priorists, the disciples of Paul Broca, chided Lombroso for his lawyerly, rather than scientific, approach. Paul Topinard said of him (1887, p. 676): "He did not say: here is a fact which suggests an induction to me, let's see if I am mistaken, let's

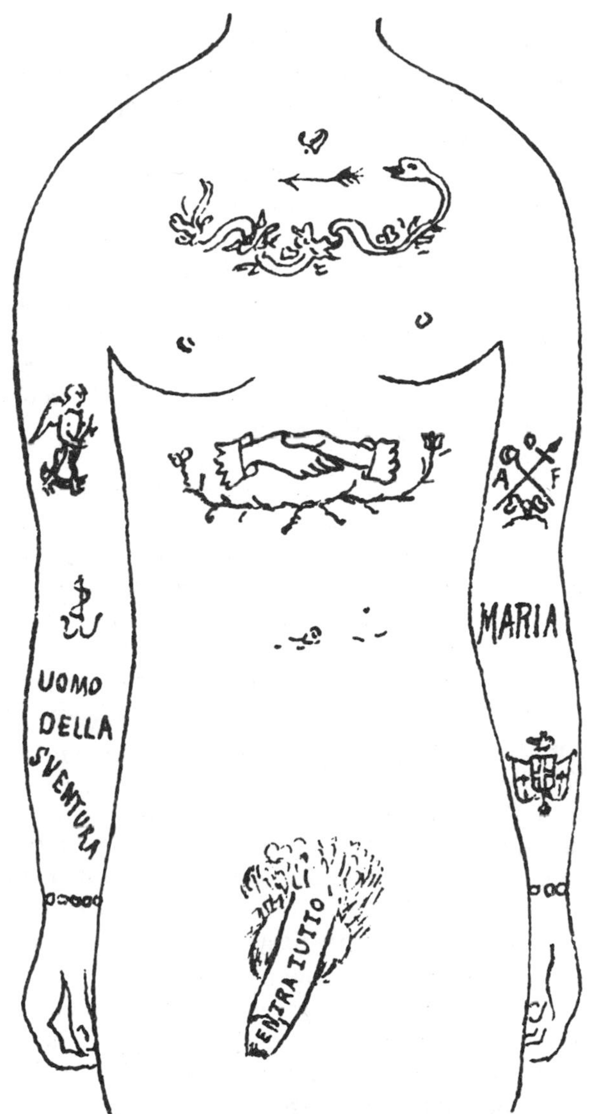

4•4 Lombroso regarded tattooing as a sign of innate criminality. The arm of this reprobate, depicted in Lombroso's *Criminal Man,* is inscribed: "A man of misfortune." On his penis we read, *entra tutto*—it enters all. In his caption, Lombroso tells us that tattoos of shaking hands are found very frequently in pederasts.

proceed rigorously, let us collect and add other facts. . . . The conclusion is fashioned in advance; he seeks proof, he defends his thesis like an advocate who ends up by persuading himself. . . . [Lombroso] is too convinced."

Lombroso slowly retreated under the barrage. But he retreated like a military master. Not for a moment did he compromise or abandon his leading idea that crime is biological. He merely enlarged the range of innate causes. His original theory had the virtue of simplicity and striking originality—criminals are apes in our midst, marked by the anatomical stigmata of atavism. Later versions became more diffuse, but also more inclusive. Atavism remained as a primary biological cause of criminal behavior, but Lombroso added several categories of congenital illness and degeneration: "We see in the criminal," he wrote (1887, p. 651), "a savage man and, at the same time, a sick man." In later years, Lombroso awarded special prominence to epilepsy as a mark of criminality; he finally stated that almost every "born criminal" suffers from epilepsy to some degree. The added burden imposed by Lombroso's theory upon thousands of epileptics cannot be calculated; they became a major target of eugenical schemes in part because Lombroso had explicated their illness as a mark of moral degeneracy.

As an intriguing sidelight, unknown to most people today, the supposed link between degeneracy and racial ranking has left us at least one legacy—the designation of "Mongolian idiocy" or, more blandly, "mongolism" for the chromosomal disorder properly known as "Down's syndrome." Dr. John Langdon Haydon Down, an English patrician, identified the syndrome in a paper entitled: "Observations on an ethnic classification of idiots" (Down, 1866).

Down argued that many congenital "idiots" (a quasi-technical term in his day, not just an epithet) exhibited anatomical features, absent in their parents but present as defining features of lower races. He found idiots of the "Ethiopian variety"—"white negroes, although of European descent" (1866, p. 260)—others of the Malay type, and "analogues of the people who with shortened foreheads, prominent cheeks, deep-set eyes, and slightly apish nose, originally inhabited the American continent" (p. 260). Others approached "the great Mongolian family." "A very large number of congenital idiots are typical Mongols" (p. 260). He then proceeded to describe, accurately, the features of Down's syndrome in a boy under his charge—a few accidental similarities with Orientals ("obliquely placed" eyes and slightly yellowish skin), and a much larger number of dissimilar features (brown and sparse hair, thick lips, wrinkled forehead, etc.). Nonetheless, he concluded (1866, p. 261): "The boy's aspect is such that it is difficult to realize that he is the child of Europeans, but so frequently are these characters presented, that there can be no doubt that these ethnic features are the result of degeneration." Down even used his ethnic insight to explain the behavior of afflicted children: "they excell at imitation"—the trait most frequently cited as typically Mongolian in conventional racist classifications of Down's time.

Down depicted himself as a racial liberal. Had he not proven human unity by showing that the characters of lower races could appear in degenerates of the higher (1866, p. 262)? In fact, he had merely done for pathology what Lombroso was soon to accomplish for criminality—to affirm the conventional racist ranks by marking undesirable whites as biological representatives of lower groups. Lombroso spoke of atavisms that "liken the European criminal to the Australian and Mongolian type" (1887, p. 254). Yet Down's designation persisted to our day and is only now fading from use. Sir Peter Medawar recently told me that he and some Oriental colleagues have recently persuaded the London *Times* to drop "mongolism" in favor of "Down's syndrome." The good doctor will still be honored.

The influence of criminal anthropology

Dallemagne, a prominent French opponent of Lombroso, paid homage to his influence in 1896:

> His thoughts revolutionized our opinions, provoked a salutary feeling everywhere, and a happy emulation in research of all kinds. For 20 years, his thoughts fed discussions; the Italian master was the order of the day in all debates; his thoughts appeared as events. There was an extraordinary animation everywhere.

Dallemagne was recording facts, not just playing diplomat. Criminal anthropology was not just an academician's debate, however lively. It was *the* subject of discussion in legal and penal circles for years. It provoked numerous "reforms" and was, until World

War I, the subject of an international conference held every four years for judges, jurists, and government officials as well as for scientists.

Beyond its specific impact, Lombrosian criminal anthropology had its primary influence in bolstering the basic argument of biological determinism about the roles of actors and their surroundings: actors follow their inborn nature. To understand crime, study the criminal, not his rearing, not his education, not the current predicament that might have inspired his theft or pillage. "Criminal anthropology studies the delinquent in his natural place—that is to say, in the field of biology and pathology" (Lombroso's disciple Sergi, quoted in Zimmern, 1898, p. 744). As a conservative political argument, it can't be beat: evil, or stupid, or poor, or disenfranchised, or degenerate, people are what they are as a result of their birth. Social institutions reflect nature. Blame (and study) the victim, not his environment.

The Italian army, for example, had been bothered by several cases of *misdeismo,* or, as we would say, fragging. The soldier Misdea (Fig. 4.5), who gave the phenomenon its Italian name, had murdered his commanding officer. Lombroso examined him and proclaimed him "a nervous epileptic . . . , very affected by a vicious heredity" (in Ferri, 1911). Lombroso recommended that epileptics be screened from the army and this, according to Ferri, eliminated *misdeismo.* (I wonder if the Italian army got through WW II without a single incident of fragging by nonepileptics.) In any case, no one seemed disposed to consider the rights and conditions of recruits.

The most dubious potential consequence of Lombroso's theory was neither realized in law nor proposed by Lombroso's supporters: prescreening and isolation of people bearing stigmata *before* they had committed any offense—though Ferri (1897, p. 251) did label as "substantially just" Plato's defense of a family's banishment after members of three successive generations had been executed for criminal offenses. Lombroso did, however, advocate prescreening of children so that teachers might prepare themselves and know what to expect from pupils with stigmata.

> Anthropological examination, by pointing out the criminal type, the precocious development of the body, the lack of symmetry, the smallness of the head, and the exaggerated size of the face explains the scholastic and disciplinary shortcomings of children thus marked and permits them

1. P. C., brigand de la Basilicate, détenu à Pesaro.

2. Voleur piémontais.

3. Incendiaire et cynède de Pesaro, surnommé *la femme*.

4. Misdea.

4•5 Four "born criminals," including the infamous Misdea, who murdered his commanding officer.

to be separated in time from their better-endowed companions and directed towards careers more suited to their temperament (1911, pp. 438–439).

We do know that Lombroso's stigmata became important criteria for judgment in many criminal trials. Again we cannot know how many men were condemned unjustly because they were extensively tattooed, failed to blush, or had unusually large jaws and arms. E. Ferri, Lombroso's chief lieutenant, wrote (1897, pp. 166–167):

> A study of the anthropological factors of crime provides the guardians and administrators of the law with new and more certain methods in the detection of the guilty. Tattooing, anthropometry, physiognomy, physical and mental conditions, records of sensibility, reflex activity, vaso-motor reactions, the range of sight, the data of criminal statistics . . . will frequently suffice to give police agents and examining magistrates a scientific guidance in their inquiries, which now depend entirely on their individual acuteness and mental sagacity. And when we remember the enormous number of crimes and offenses which are not punished, for lack or inadequacy of evidence, and the frequency of trials which are based solely on circumstantial hints, it is easy to see the practical utility of the primary connection between criminal sociology and penal procedure.

Lombroso detailed some of his experiences as an expert witness. Called upon to help decide which of two stepsons had killed a woman, Lombroso declared (1911, p. 436) that one "was, in fact, the most perfect type of the born criminal: enormous jaws, frontal sinuses, and zygomata, thin upper lip, huge incisors, unusually large head (1620 cc) [a mark of genius in other contexts, but not here], tactile obtuseness with sensorial manicinism. He was convicted."

In another case, based on evidence that even he could not depict as better than highly vague and circumstantial, Lombroso argued for the conviction of a certain Fazio, accused of robbing and murdering a rich farmer. One girl testified that she had seen Fazio sleeping near the murdered man; the next morning he hid as the gendarmes approached. No other evidence of his guilt was offered:

> Upon examination I found that this man had outstanding ears, great maxillaries and cheek bones, lemurine appendix, division of the frontal bone, premature wrinkles, sinister look, nose twisted to the right—in short, a physiognomy approaching the criminal type; pupils very slightly mobile

> . . . a large picture of a woman tattooed upon his breast, with the words, "Remembrance of Celina Laura" (his wife), and on his arm the picture of a girl. He had an epileptic aunt and an insane cousin, and investigation showed that he was a gambler and an idler. In every way, then, biology furnished in this case indications which, joined with the other evidence, would have been enough to convict him in a country less tender toward criminals. Notwithstanding this he was acquitted (1911, p. 437).

You win some, you lose some. (Ironically, it was the conservative rather than the liberal nature of jurisprudence that limited Lombroso's influence. Most judges and lawyers simply couldn't bear the idea of quantitative science intruding into their ancient domain. They didn't know that Lombrosian criminal anthropology was a pseudo-science, but rejected it as an unwarranted transgression of a study fully *legitimate* in its own domain. Lombroso's French critics, with their emphasis on the social causes of crime, also helped to halt the Lombrosian tide—for they, Manouvrier and Topinard in particular, could parry numbers with him.)

In discussing capital punishment, Lombroso and his disciples emphasized their conviction that born criminals transgress by nature. "Atavism shows us the inefficacy of punishment for born criminals and why it is that they inevitably have periodic relapses into crime" (Lombroso, 1911, p. 369). "Theoretical ethics passes over these diseased brains, as oil does over marble, without penetrating it" (Lombroso, 1895, p. 58).

Ferri stated in 1897 that, in opposition to many other schools of thought, criminal anthropologists of Lombrosian persuasion were unanimous in declaring the death penalty legitimate (1897, pp. 238–240). Lombroso wrote (1911, p. 447): "There exists, it is true, a group of criminals, born for evil, against whom all social cures break as against a rock—a fact which compels us to eliminate them completely, even by death." His friend the philosopher Hippolyte Taine wrote even more dramatically:

> You have shown us fierce and lubricious orang-utans with human faces. It is evident that as such they cannot act otherwise. If they ravish, steal, and kill, it is by virtue of their own nature and their past, but there is all the more reason for destroying them when it has been proved that they will always remain orang-utans (quoted favorably in Lombroso, 1911, p. 428).

Ferri himself invoked Darwinian theory as a cosmic justification for capital punishment (1897, pp. 239–240):

It seems to me that the death penalty is prescribed by nature, and operates at every moment in the life of the universe. The universal law of evolution shows us also that vital progress of every kind is due to continual selection, by the death of the least fit in the struggle for life. Now this selection, in humanity as with the lower animals, may be natural or artificial. It would therefore be in agreement with natural laws that human society should make an artificial selection, by the elimination of anti-social and incongruous individuals.

Nonetheless, Lombroso and his colleagues generally favored means other than death for ridding society of its born criminals. Early isolation in bucolic surroundings might mitigate the innate tendency and lead to a useful life under close and continual supervision. In other cases of incorrigible criminality, transportation and exile to penal colonies provided a more humanitarian solution than capital punishment—but banishment must be permanent and irrevocable. Ferri, noting the small size of Italy's colonial empire, advocated "internal deportation," perhaps to lands not tilled because of endemic malaria: "If the dispersion of this malaria demands a human hecatomb, it would evidently be better to sacrifice criminals than honest husbandmen" (1897, p. 249). In the end, he recommended deportation to the African colony of Eritrea.

The Lombrosian criminal anthropologists were not petty sadists, proto-fascists, or even conservative political ideologues. They tended toward liberal, even socialist, politics and saw themselves as scientifically enlightened modernists. They hoped to use modern science as a cleansing broom to sweep away from jurisprudence the outdated philosophical baggage of free will and unmitigated moral responsibility. They called themselves the "positive" school of criminology, not because they were so certain (though they were), but in reference to the philosophical meaning of empirical and objective rather than speculative.

The "classical" school, Lombroso's chief opponents, had combatted the capriciousness of previous penal practice by arguing that punishment must be apportioned strictly to the nature of the crime and that all individuals must be fully responsible for their actions (no mitigating circumstances). Lombroso invoked biology to argue that punishments must fit the criminal, not, as Gilbert's Mikado would have it, the crime. A normal man might murder in a moment of jealous rage. What purpose would execution or a life in prison serve? He needs no reform, for his nature is good; society needs no protection from him, for he will not transgress again. A born criminal might be in the dock for some petty crime. What good will a short sentence serve: since he cannot be rehabilitated, a short sentence only reduces the time to his next, perhaps more serious, offense.

The positive school campaigned hardest and most successfully for a set of reforms, until recently regarded as enlightened or "liberal," and all involving the principle of indeterminate sentencing. For the most part they won, and few people realize that our modern apparatus of parole, early release, and indeterminate sentencing stems in part from Lombroso's campaign for differential treatment of born and occasional criminals. The main goal of criminal anthropology, wrote Ferri in 1911, is to "make the personality of the criminal the primary object and principle of the rules of penal justice, in place of the objective gravity of the crime" (p. 52).

> Penal sanctions must be adapted . . . to the personality of the criminal. . . . The logical consequence of this conclusion is the indeterminate sentence which has been, and is, combatted as a juridical heresy by classical and metaphysical criminologists. . . . Prefixed penalties are absurd as a means of social defense. It is as if a doctor at the hospital wanted to attach to each disease the length of a stay in his establishment (Ferri, 1911, p. 251).

The original Lombrosians advocated harsh treatment for "born criminals." This misapplication of anthropometry and evolutionary theory is all the more tragic because Lombroso's biological model was so utterly invalid and because it shifted so much attention from the social basis of crime to fallacious ideas about the innate propensity of criminals. But the positivists, invoking Lombroso's enlarged model and finally even extending the genesis of crime to upbringing as well as biology, had enormous impact in their campaign for indeterminate sentencing and the concept of mitigating circumstances. Since their beliefs are, for the most part, our practices, we have tended to view them as humane and progressive. Lombroso's daughter, carrying on the good work, singled out the United States for praise. We had escaped the hegemony of classical criminology and shown our usual receptiveness for innovation. Many states had adopted the positivist program in establishing good reformatories, probation systems, indeterminate sentencing, and liberal pardon laws (Lombroso-Ferrero, 1911).

Yet even as the positivists praised America and themselves,

their work contains the seeds of doubt that have led many modern reformers to question the humane nature of Lombroso's indeterminate sentences and to advocate a return to the fixed penalties of classical criminology. Maurice Parmelee, America's leading positivist, decried as too harsh a New York State law of 1915 that prescribed an indeterminate sentence of up to three years for such infractions as disorderly conduct, disorderly housekeeping, intoxication, and vagrancy (Parmelee, 1918). Lombroso's daughter praised the complete dossier of moods and deeds kept by volunteer women who guided the fortunes of juvenile offenders in several states. They will "permit judges, if the child commits an offense, to distinguish between a born criminal and a habitual criminal. However, the child will not know of the existence of this dossier, and this will permit him the most complete freedom to develop" Lombroso-Ferrero, 1911, p. 124). She also admitted the burdensome element of harassment and humiliation included in several systems of probation, particularly in Massachusetts, where indefinite parole might continue for life: "In the Central Probation Bureau of Boston, I have read many letters from protégés who asked to be returned to their prisons, rather than continue the humiliation of their protector always on their backs (or "in their bundles," as she said literally in French—Lombroso-Ferrero, 1911, p. 135).

For the Lombrosians, indeterminate sentencing embodied both good biology and maximal protection for the state: "Punishment ought not to be the visitation of a crime by a retribution, but rather a defense of society adapted to the danger personified by the criminal" (Ferri, 1897, p. 208). Dangerous people receive longer sentences, and their subsequent lives are monitored more strictly. And so the system of indeterminate penalties—Lombroso's legacy—exerts a general and powerful element of control over every aspect of a prisoner's life: his dossier expands and controls his fate; he is watched in prison and his acts are judged with the carrot of early release before him. It is also used in Lombroso's original sense to sequester the dangerous. For Lombroso, this meant the born criminal with his apish stigmata. Today, it often means the defiant, the poor, and the black. George Jackson, author of *Soledad Brother,* died under Lombroso's legacy, trying to escape after eleven years (eight and a half in solitary) of an indeterminate one-year-to-life sentence for stealing seventy dollars from a gas station.

Coda

Tolstoy's frustration with the Lombrosians lay in their invocation of science to avoid the deeper question that called for social transformation as one potential resolution. Science, he realized, often acted as the firm ally of existing institutions. His protagonist Prince Nekhlyudov, trying to fathom a system that falsely condemned a woman he once wronged, studies the learned works of criminal anthropology and finds no answer:

> He also came across a tramp and a woman, both of whom repelled him by their half-witted insensibility and seeming cruelty, but even in them he failed to see the criminal type as described in the Italian school of criminology: he saw in them only people who were repulsive to him personally, like others were whom he met outside prison walls—in swallowtail coats, wearing epaulets or bedecked with lace. . . .
>
> At first he had hoped to find the answer in books, and bought everything he could find on the subject. He bought the works of Lombroso and Garofalo [an Italian baron and disciple of Lombroso], Ferri, Liszt, Maudsley and Tarde, and read them carefully. But as he read, he became more and more disappointed. . . . Science answered thousands of very subtle and ingenious questions touching criminal law, but certainly not the one he was trying to solve. He was asking a very simple thing: Why and by what right does one class of people lock up, torture, exile, flog, and kill other people, when they themselves are no better than those whom they torture, flog and kill? And for answers he got arguments as to whether human beings were possessed of free will or not. Could criminal propensities be detected by measuring the skull, and so on? What part does heredity play in crime? Is there such a thing as congenital depravity? (*Resurrection,* 1899, 1966 edition translated by R. Edmonds, pp. 402–403.)

Epilogue

We live in a more subtle century, but the basic arguments never seem to change. The crudities of the cranial index have given way to the complexity of intelligence testing. The signs of innate criminality are no longer sought in stigmata of gross anatomy, but in twentieth-century criteria: genes and fine structure of the brain.

In the mid-1960s, papers began to appear linking a chromosomal anomaly in males known as XYY with violent and criminal behavior. (Normal males receive a single X chromosome from their mothers and a Y from their fathers; normal females receive a single X from each of their parents. Occasionally, a child will receive two Y's from his father. XYY males look like normal males, but tend to be a little above average in height, have poor skin and may tend, on average—though this is disputed—to be somewhat deficient in performance on intelligence tests.) Based on limited observation and anecdotal accounts of a few XYY individuals, and on a high frequency of XYY's in mental-penal institutions for the criminally insane, a tale about criminal chromosomes originated. The story exploded into public consciousness when attorneys for Richard Speck, murderer of eight student nurses in Chicago, sought to mitigate his punishment with a claim that he was XYY. (In fact, he is a normal XY male.) *Newsweek* published an article entitled "Congenital criminals," and the press churned out innumerable reports about this latest reincarnation of Lombroso and his stigmata. Meanwhile, scholarly study picked up, and hundreds of papers have now been written on the behavioral consequences of being XYY. A well-intentioned but, in my opinion, naïve group of Boston doctors began an extensive screening program upon newborn boys. They hoped that by monitoring the development of a large sample of XYY boys, they might establish whether any link existed with aggressive behavior. But what of the self-fulfilling prophesy? for parents were told, and no amount of scholarly tentativeness can overcome both press reports and inferences made by worried parents from the aggressive behavior manifested from time to time by all children. And what of the anguish suffered by parents, especially if the connection be a false one—as it almost surely is.

In theory, the link between XYY and aggressive criminality never had much going for it beyond the singularly simplistic notion that since males are more aggressive than females and possess a Y that females lack, Y must be the seat of aggression and a double dose spells double-trouble. One group of researchers proclaimed in 1973 (Jarvik et al., pp. 679–680): "The Y chromosome is the male-determining chromosome; therefore, it should come as no surprise that an extra Y chromosome can produce an individual with heightened masculinity, evinced by characteristics such as unusual tallness, increased fertility . . . and powerful aggressive tendencies."

The tale of XYY as a criminal stigma has now been widely exposed as a myth (Borgaonkar and Shah, 1974; Pyeritz et al.,

1977). Both these studies expose the elementary flaws of method in most literature claiming a link between XYY and criminality. XYY males do seem to be represented disproportionately in mental-penal institutions, but there is no good evidence for high frequencies in ordinary jails. A maximum of 1 percent of XYY males in America may spend part of their lives in mental-penal institutions (Pyeritz et al., 1977, p. 92). Adding to this the number that may be incarcerated in ordinary jails at the same frequency as normal XY males, Chorover (1979) estimates that 96 percent of XYY males will lead ordinary lives and never come to the attention of penal authorities. Quite a criminal chromosome! Moreover, we have no evidence that the relatively high proportion of XYY's in mental-penal institutions has anything to do with high levels of innate aggressivity.

Other scientists have looked to malfunction in specific areas of the brain as a cause of criminal behavior. After extensive ghetto riots during the summer of 1967, three doctors wrote a letter to the prestigious *Journal of the American Medical Association* (cited in Chorover, 1979):

> It is important to realize that only a small number of the millions of slum dwellers have taken part in the riots, and that only a subfraction of these rioters have indulged in arson, sniping and assault. Yet, if slum conditions alone determined and initiated riots, why are the vast majority of slum dwellers able to resist the temptations of unrestrained violence? Is there something peculiar about the violent slum dweller that differentiates him from his peaceful neighbors?

We all tend to generalize from our own areas of expertise. These doctors are psychosurgeons. But why should the violent behavior of some desperate and discouraged people point to a specific disorder of their brain while the corruption and violence of some congressmen and presidents provokes no similar theory? Human populations are highly variable for all behaviors; the simple fact that some do and some don't provides no evidence for a specific pathology mapped upon the brain of doers. Shall we concentrate upon an unfounded speculation for the violence of some—one that follows the determinist philosophy of blaming the victim—or shall we try to eliminate the oppression that builds ghettos and saps the spirit of their unemployed in the first place?

FIVE

The Hereditarian Theory of IQ

An American Invention

Alfred Binet and the original purposes of the Binet scale

Binet flirts with craniometry

When Alfred Binet (1857–1911), director of the psychology laboratory at the Sorbonne, first decided to study the measurement of intelligence, he turned naturally to the favored method of a waning century and to the work of his great countryman Paul Broca. He set out, in short, to measure skulls, never doubting at first the basic conclusion of Broca's school:

> The relationship between the intelligence of subjects and the volume of their head . . . is very real and has been confirmed by all methodical investigators, without exception. . . . As these works include observations on several hundred subjects, we conclude that the preceding proposition [of correlation between head size and intelligence] must be considered as incontestable (Binet, 1898, pp. 294–295).

During the next three years, Binet published nine papers on craniometry in *L'Année psychologique,* the journal he had founded in 1895. By the end of this effort, he was no longer so sure. Five studies on the heads of school children had destroyed his original faith.

Binet went to various schools, making Broca's recommended measurements on the heads of pupils designated by teachers as their smartest and stupidest. In several studies, he increased his sample from 62 to 230 subjects. "I began," he wrote, "with the idea, impressed upon me by the studies of so many other scientists, that intellectual superiority is tied to superiority of cerebral volume" (1900, p. 427).

Binet found his differences, but they were much too small to matter and might only record the greater average height of better pupils (1.401 vs. 1.378 meters). Most measures did favor the better students, but the average difference between good and poor amounted to a mere millimeter—*"extrêmement petite"* as Binet wrote. Binet did not observe larger differences in the anterior region of the skull, where the seat of higher intelligence supposedly lay, and where Broca had always found greatest disparity between superior and less fortunate people. To make matters worse, some measures usually judged crucial in the assessment of mental worth favored the poorer pupils—for anteroposterior diameter of the skull, poorer students exceeded their smarter colleagues by 3.0 mm. Even if most results tended to run in the "right" direction, the method was surely useless for assessing individuals. The differences were too small, and Binet also found that poor students varied more than their smarter counterparts. Thus, although the smallest value usually belonged to a poor pupil, the highest often did as well.

Binet also fueled his own doubts with an extraordinary study of his own suggestibility, an experiment in the primary theme of this book—the tenacity of unconscious bias and the surprising malleability of "objective," quantitative data in the interest of a preconceived idea. "I feared," Binet wrote (1900, p. 323), "that in making measurements on heads with the intention of finding a difference in volume between an intelligent and a less intelligent head, I would be led to increase, unconsciously and in good faith, the cephalic volume of intelligent heads and to decrease that of unintelligent heads." He recognized the greater danger lurking when biases are submerged and a scientist believes in his own objectivity (1900, p. 324): "Suggestibility . . . works less on an act of which we have full consciousness, than on a half-conscious act—and this is precisely its danger."

How much better off we would be if all scientists submitted themselves to self-scrutiny in so forthright a fashion: "I want to state very explicitly," Binet wrote (1900, p. 324), "what I have observed about myself. The details that follow are those that the majority of authors do not publish; one does not want to let them be known." Both Binet and his student Simon had measured the same heads of "idiots and imbeciles" at a hospital where Simon was in intern. Binet noted that, for one crucial measurement, Simon's values were consistently less than his. Binet therefore returned to measure the subjects a second time. The first time, Binet admits, "I took my measures mechanically, without any other preconception than to remain faithful to my methods." But the second time "I had a different preconception. . . . I was bothered by the difference" between Simon and myself. "I wanted to reduce it to its true value. . . . This is self-suggestion. Now, capital fact, the measures taken during the second experiment, under the expectation of a diminution, are indeed smaller than the measures taken [on the same heads] during the first experiment." In fact, all but one head had "shrunk" between the two experiments and the average diminution was 3 mm—a good deal more than the average difference between skulls of bright and poor students in his previous work.

Binet spoke graphically of his discouragement:

> I was persuaded that I had attacked an intractable problem. The measures had required travelling, and tiring procedures of all sorts; and they ended with the discouraging conclusion that there was often not a millimeter of difference between the cephalic measures of intelligent and less intelligent students. The idea of measuring intelligence by measuring heads seemed ridiculous. . . . I was on the point of abandoning this work, and I didn't want to publish a single line of it (1900, p. 403).

At the end, Binet snatched a weak and dubious victory from the jaws of defeat. He looked at his entire sample again, separated out the five top and bottom pupils from each group, and eliminated all those in the middle. The differences between extremes were greater and more consistent—3 to 4 mm on average. But even this difference did not exceed the average potential bias due to suggestibility. Craniometry, the jewel of nineteenth-century objectivity, was not destined for continued celebration.

Binet's scale and the birth of IQ

When Binet returned to the measurement of intelligence in 1904, he remembered his previous frustration and switched to other techniques. He abandoned what he called the "medical" approaches of craniometry and the search for Lombroso's anatomical stigmata, and decided instead on "psychological" methods. The literature on mental testing, at the time, was relatively small and decidedly inconclusive. Galton, without notable success, had experimented with a series of measurements, mostly records of physiology and reaction time, rather than tests of reasoning. Binet decided to construct a set of tasks that might assess various aspects of reasoning more directly.

In 1904 Binet was commissioned by the minister of public education to perform a study for a specific, practical purpose: to develop techniques for identifying those children whose lack of success in normal classrooms suggested the need for some form of special education. Binet chose a purely pragmatic course. He decided to bring together a large series of short tasks, related to everyday problems of life (counting coins, or assessing which face is "prettier," for example), but supposedly involving such basic processes of reasoning as "direction (ordering), comprehension, invention and censure (correction)" (Binet, 1909). Learned skills like reading would not be treated explicitly. The tests were administered individually by trained examiners who led subjects through the series of tasks, graded in their order of difficulty. Unlike previous tests designed to measure specific and independent "faculties" of mind, Binet's scale was a hodgepodge of diverse activities. He hoped that by mixing together enough tests of different abilities he would be able to abstract a child's general potential with a single score. Binet emphasized the empirical nature of his work with a famous dictum (1911, p. 329): "One might almost say, 'It matters very little what the tests are so long as they are numerous.' "

Binet published three versions of the scale before his death in 1911. The original 1905 edition simply arranged the tasks in an ascending order of difficulty. The 1908 version established the criterion used in measuring the so-called IQ ever since. Binet decided to assign an age level to each task, defined as the youngest age at which a child of normal intelligence should be able to complete the task successfully. A child began the Binet test with tasks for the youngest age and proceeded in sequence until he could no longer complete the tasks. The age associated with the last tasks he could perform became his "mental age," and his general intellectual level was calculated by subtracting this mental age from his true chronological age. Children whose mental ages were sufficiently behind their chronological ages could then be identified for special educational programs, thus fulfilling Binet's charge from the ministry. In 1912 the German psychologist W. Stern argued that mental age should be divided by chronological age, not subtracted from it,* and the intelligence *quotient,* or IQ, was born.

IQ testing has had momentous consequences in our century. In this light, we should investigate Binet's motives, if only to appreciate how the tragedies of misuse might have been avoided if its founder had lived and his concerns been heeded.

In contrast with Binet's general intellectual approach, the most curious aspect of his scale is its practical, empirical focus. Many scientists work this way by deep conviction or explicit inclination. They believe that theoretical speculation is vain and that true science progresses by induction from simple experiments pursued to gather basic facts, not to test elaborate theories. But Binet was primarily a theoretician. He asked big questions and participated with enthusiasm in the major philosophical debates of his profession. He had a long-standing interest in theories of intelligence. He published his first book on the "Psychology of Reasoning" in 1886, and followed in 1903 with his famous "Experimental Study of Intelligence," in which he abjured previous commitments and developed a new structure for analyzing human thinking. Yet Binet explicitly declined to award any theoretical interpretation to his scale of intelligence, the most extensive and important work he had done in his favorite subject. Why should a great theoretician have acted in such a curious and apparently contradictory way?

Binet did seek "to separate natural intelligence and instruction" (1905, p. 42) in his scale: "It is the intelligence alone that we seek to measure, by disregarding in so far as possible, the degree of instruction which the child possesses. . . . We give him nothing to read, nothing to write, and submit him to no test in which he might

*Division is more appropriate because it is the relative, not the absolute, magnitude of disparity between mental and chronological age that matters. A two-year disparity between mental age two and chronological age four may denote a far severer degree of deficiency than a two-year disparity between mental age fourteen and chronological age sixteen. Binet's method of subtraction would give the same result in both cases, while Stern's IQ measures 50 for the first case and 88 for the second. (Stern multiplied the actual quotient by 100 to eliminate the decimal point.)

succeed by means of rote learning" (1905, p. 42). "It is a specially interesting feature of these tests that they permit us, when necessary, to free a beautiful native intelligence from the trammels of the school" (1908, p. 259).

Yet, beyond this obvious desire to remove the superficial effects of clearly acquired knowledge, Binet declined to define and speculate upon the meaning of the score he assigned to each child. Intelligence, Binet proclaimed, is too complex to capture with a single number. This number, later called IQ, is only a rough, empirical guide constructed for a limited, practical purpose:

> The scale, properly speaking, does not permit the measure of the intelligence, because intellectual qualities are not superposable, and therefore cannot be measured as linear surfaces are measured (1905, p. 40).

Moreover, the number is only an average of many performances, not an entity unto itself. Intelligence, Binet reminds us, is not a single, scalable thing like height. "We feel it necessary to insist on this fact," Binet (1911) cautions, "because later, for the sake of simplicity of statement, we will speak of a child of 8 years having the intelligence of a child of 7 or 9 years; these expressions, if accepted arbitrarily, may give place to illusions." Binet was too good a theoretician to fall into the logical error that John Stuart Mill had identified—"to believe that whatever received a name must be an entity or being, having an independent existence of its own."

Binet also had a social motive for his reticence. He greatly feared that his practical device, if reified as an entity, could be perverted and used as an indelible label, rather than as a guide for identifying children who needed help. He worried that schoolmasters with "exaggerated zeal" might use IQ as a convenient excuse: "They seem to reason in the following way: 'Here is an excellent opportunity for getting rid of all the children who trouble us,' and without the true critical spirit, they designate all who are unruly, or disinterested in the school" (1905, p. 169). But he feared even more what has since been called the "self-fulfilling prophesy." A rigid label may set a teacher's attitude and eventually divert a child's behavior into a predicted path:

> It is really too easy to discover signs of backwardness in an individual when one is forewarned. This would be to operate as the graphologists did

who, when Dreyfus was believed to be guilty, discovered in his handwriting signs of a traitor or a spy" (1905, p. 170).

Not only did Binet decline to label IQ as inborn intelligence; he also refused to regard it as a general device for ranking all pupils according to mental worth. He devised his scale only for the limited purpose of his commission by the ministry of education: as a practical guide for identifying children whose poor performance indicated a need for special education—those who we would today call learning disabled or mildly retarded. Binet wrote (1908, p. 263): "We are of the opinion that the most valuable use of our scale will not be its application to the normal pupils, but rather to those of inferior grades of intelligence." As to the causes of poor performance, Binet refused to speculate. His tests, in any case, could not decide (1905, p. 37):

> Our purpose is to be able to measure the intellectual capacity of a child who is brought to us in order to know whether he is normal or retarded. We should therefore study his condition at the time and that only. We have nothing to do either with his past history or with his future; consequently, we shall neglect his etiology, and we shall make no attempt to distinguish between acquired and congenital idiocy. . . . As to that which concerns his future, we shall exercise the same abstinence; we do not attempt to establish or prepare a prognosis, and we leave unanswered the question of whether this retardation is curable, or even improvable. We shall limit ourselves to ascertaining the truth in regard to his present mental state.

But of one thing Binet was sure: whatever the cause of poor performance in school, the aim of his scale was to identify in order to help and improve, not to label in order to limit. Some children might be innately incapable of normal achievement, but all could improve with special help.

The difference between strict hereditarians and their opponents is not, as some caricatures suggest, the belief that a child's performance is all inborn or all a function of environment and learning. I doubt that the most committed antihereditarians have ever denied the existence of innate variation among children. The differences are more a matter of social policy and educational practice. Hereditarians view their measures of intelligence as markers of permanent, inborn limits. Children, so labeled, should be sorted, trained according to their inheritance and channeled into professions appropriate for their biology. Mental testing becomes a theory of limits. Antihereditarians, like Binet, test in order to identify and help. Without denying the evident fact that not all children, whatever their training, will enter the company of Newton and Einstein, they emphasize the power of creative education to increase the achievements of all children, often in extensive and unanticipated ways. Mental testing becomes a theory for enhancing potential through proper education.

Binet spoke eloquently of well-meaning teachers, caught in the unwarranted pessimism of their invalid hereditarian assumptions (1909, pp. 16–17):

> As I know from experience, . . . they seem to admit implicitly that in a class where we find the best, we must also find the worst, and that this is a natural and inevitable phenomenon, with which a teacher must not become preoccupied, and that it is like the existence of rich and poor within a society. What a profound error.

How can we help a child if we label him as unable to achieve by biological proclamation?

> If we do nothing, if we don't intervene actively and usefully, he will continue to lose time . . . and will finally become discouraged. The situation is very serious for him, and since his is not an exceptional case (since children with defective comprehension are legion), we might say that it is a serious question for all of us and for all of society. The child who loses the taste for work in class strongly risks being unable to acquire it after he leaves school (1909, p. 100).

Binet railed against the motto "stupidity is for a long time" ("*quand on est bête, c'est pour longtemps*"), and upbraided teachers who "are not interested in students who lack intelligence. They have neither sympathy nor respect for them, and their intemperate language leads them to say such things in their presence as 'This is a child who will never amount to anything . . . he is poorly endowed . . . he is not intelligent at all.' How often have I heard these imprudent words" (1909, p. 100). Binet then cites an episode in his own baccalaureate when one examiner told him that he would never have a "true" philosophical spirit: "Never! What a momentous word. Some recent thinkers seem to have given their moral support to these deplorable verdicts by affirming that an individual's intelligence is a fixed quantity, a quantity that cannot be increased. We must protest and react against this brutal pessimism; we must try to demonstrate that it is founded upon nothing" (1909, p. 101).

The children identified by Binet's test were to be helped, not indelibly labeled. Binet had definite pedagogical suggestions, and many were implemented. He believed, first of all, that special education must be tailored to the individual needs of disadvantaged children: it must be based on "their character and their aptitudes, and on the necessity for adapting ourselves to their needs and their capacities" (1909, p. 15). Binet recommended small classrooms of fifteen to twenty students, compared with sixty to eighty then common in public schools catering to poor children. In particular, he advocated special methods of education, including a program that he called "mental orthopedics":

> What they should learn first is not the subjects ordinarily taught, however important they may be; they should be given lessons of will, of attention, of discipline; before exercises in grammar, they need to be exercised in mental orthopedics; in a word they must learn how to learn (1908, p. 257).

Binet's interesting program of mental orthopedics included a set of physical exercises designed to improve, by transfer to mental functioning, the will, attention, and discipline that Binet viewed as prerequisites for studying academic subjects. In one, called "*l'exercise des statues,*" and designed to increase attention span, children moved vigorously until told to adopt and retain an immobile position. (I played this game as a kid in the streets of New York; we also called it "statues.") Each day the period of immobility would be increased. In another, designed to improve speed, children filled a piece of paper with as many dots as they could produce in the allotted time.

Binet spoke with pleasure about the success of his special classrooms (1909, p. 104) and argued that pupils so benefited had not only increased their knowledge, but their intelligence as well. Intelligence, in any meaningful sense of the word, can be augmented by good education; it is not a fixed and inborn quantity:

> It is in this practical sense, the only one accessible to us, that we say that the intelligence of these children has been increased. We have increased what constitutes the intelligence of a pupil: the capacity to learn and to assimilate instruction.

The dismantling of Binet's intentions in America

In summary, Binet insisted upon three cardinal principles for using his tests. All his caveats were later disregarded, and his intentions overturned, by the American hereditarians who translated his scale into written form as a routine device for testing all children.

1. The scores are a practical device; they do not buttress any theory of intellect. They do not define anything innate or permanent. We may not designate what they measure as "intelligence" or any other reified entity.

2. The scale is a rough, empirical guide for identifying mildly retarded and learning-disabled children who need special help. It is not a device for ranking normal children.

3. Whatever the cause of difficulty in children identified for help, emphasis shall be placed upon improvement through special training. Low scores shall not be used to mark children as innately incapable.

If Binet's principles had been followed, and his tests consistently used as he intended, we would have been spared a major misuse of science in our century. Ironically, many American school boards have come full cycle, and now use IQ tests only as Binet originally recommended: as instruments for assessing children with specific learning problems. Speaking personally, I feel that tests of the IQ type were helpful in the proper diagnosis of my own learning-disabled son. His average score, the IQ itself, meant nothing, for it was only an amalgam of some very high and very low scores; but the pattern of low values indicated his areas of deficit.

The misuse of mental tests is not inherent in the idea of testing itself. It arises primarily from two fallacies, eagerly (so it seems) embraced by those who wish to use tests for the maintenance of social ranks and distinctions: reification and hereditarianism. The next chapter shall treat reification—the assumption that test scores represent a single, scalable thing in the head called general intelligence.

The hereditarian fallacy is not the simple claim that IQ is to some degree "heritable." I have no doubt that it is, though the degree has clearly been exaggerated by the most avid hereditarians. It is hard to find any broad aspect of human performance or anatomy that has no heritable component at all. The hereditarian fallacy resides in two false implications drawn from this basic fact:

1. The equation of "heritable" with "inevitable." To a biologist, heritability refers to the passage of traits or tendencies along family lines as a result of genetic transmission. It says little about the range of environmental modification to which these traits are subject. In our vernacular, "inherited" often means "inevitable." But not to a biologist. Genes do not make specific bits and pieces of a body; they code for a range of forms under an array of environmental conditions. Moreover, even when a trait has been built and set, environmental intervention may still modify inherited defects. Millions of Americans see normally through lenses that correct innate deficiencies of vision. The claim that IQ is so-many percent "heritable" does not conflict with the belief that enriched education can increase what we call, also in the vernacular, "intelligence." A partially inherited low IQ might be subject to extensive improvement through proper education. And it might not. The mere fact of its heritability permits no conclusion.

2. The confusion of within- and between-group heredity. The major political impact of hereditarian theories does not arise from the inferred heritability of tests, but from a logically invalid extension. Studies of the heritability of IQ, performed by such traditional methods as comparing scores of relatives, or contrasting scores of adopted children with both their biological and legal parents, are all of the "within-group" type—that is, they permit an estimate of heritability *within* a single, coherent population (white Americans, for example). The common fallacy consists in assuming that if heredity explains a certain percentage of variation among individuals within a group, it must also explain a similar percentage of the difference in average IQ between groups—whites and blacks, for example. But variation among individuals within a group and differences in mean values between groups are entirely separate phenomena. One item provides no license for speculation about the other.

A hypothetical and noncontroversial example will suffice. Human height has a higher heritability than any value ever proposed for IQ. Take two separate groups of males. The first, with an average height of 5 feet 10 inches, live in a prosperous American town. The second, with an average height of 5 feet 6 inches, are starving in a third-world village. Heritability is 95 percent or so in each place—meaning only that relatively tall fathers tend to have tall sons and relatively short fathers short sons. This high within-group heritability argues neither for nor against the possibility that better nutrition in the next generation might raise the average height of third-world villagers above that of prosperous Americans. Likewise, IQ could be highly heritable within groups, and the average difference between whites and blacks in America might still only record the environmental disadvantages of blacks.

I have often been frustrated with the following response to this admonition: "Oh well, I see what you mean, and you're right in theory. There may be no necessary connection in logic, but isn't it more likely all the same that mean differences between groups would have the same causes as variation within groups." The answer is still "no." Within- and between-group heredity are not tied by rising degrees of probability as heritability increases within groups and differences enlarge between them. The two phenomena are simply separate. Few arguments are more dangerous than the ones that "feel" right but can't be justified.

Alfred Binet avoided these fallacies and stuck by his three principles. American psychologists perverted Binet's intention and invented the hereditarian theory of IQ. They reified Binet's scores, and took them as measures of an entity called intelligence. They assumed that intelligence was largely inherited, and developed a series of specious arguments confusing cultural differences with innate properties. They believed that inherited IQ scores marked people and groups for an inevitable station in life. And they assumed that average differences between groups were largely the products of heredity, despite manifest and profound variation in quality of life.

This chapter analyzes the major works of the three pioneers of hereditarianism in America: H. H. Goddard, who brought Binet's scale to America and reified its scores as innate intelligence; L. M. Terman, who developed the Stanford-Binet scale, and dreamed of a rational society that would allocate professions by IQ scores; and R. M. Yerkes, who persuaded the army to test 1.75 million men in World War I, thus establishing the supposedly objective data that vindicated hereditarian claims and led to the Immigration Restriction Act of 1924, with its low ceiling for lands suffering the blight of poor genes.

The hereditarian theory of IQ is a home-grown American

product. If this claim seems paradoxical for a land with egalitarian traditions, remember also the jingoistic nationalism of World War I, the fear of established old Americans facing a tide of cheap (and sometimes politically radical) labor immigrating from southern and eastern Europe, and above all our persistent, indiginous racism.

H. H. Goddard and the menace of the feeble-minded

Intelligence as a Mendelian gene

GODDARD IDENTIFIES THE MORON

It remains now for someone to determine the nature of feeble-mindedness and complete the theory of the intelligence quotient.

—H. H. GODDARD, 1917, in a review of Terman, 1916

Taxonomy is always a contentious issue because the world does not come to us in neat little packages. The classification of mental deficiency aroused a healthy debate early in our century. Two categories of a tripartite arrangement won general acceptance: idiots could not develop full speech and had mental ages below three; imbeciles could not master written language and ranged from three to seven in mental age. (Both terms are now so entrenched in the vernacular of invectives that few people recognize their technical status in an older psychology.) Idiots and imbeciles could be categorized and separated to the satisfaction of most professionals, for their affliction was sufficiently severe to warrant a diagnosis of true pathology. They are not like us.

But consider the nebulous and more threatening realm of "high-grade defectives"—the people who could be trained to function in society, the ones who established a bridge between pathology and normality and thereby threatened the taxonomic edifice. These people, with mental ages of eight to twelve, were called *débile* (or weak) by the French. Americans and Englishmen usually called them "feeble-minded," a term mired in hopeless ambiguity because other psychologists used feeble-minded as a generic term for all mental defectives, not just those of high grade.

Taxonomists often confuse the invention of a name with the solution of a problem. H. H. Goddard, the energetic and crusading director of research at the Vineland Training School for Feeble-Minded Girls and Boys in New Jersey, made this crucial error. He devised a name for "high-grade" defectives, a word that became entrenched in our language through a series of jokes that rivaled the knock-knock or elephant jokes of other generations. The metaphorical whiskers on these jokes are now so long that most people would probably grant an ancient pedigree to the name. But Goddard invented the word in our century. He christened these people "morons," from a Greek word meaning foolish.

Goddard was the first popularizer of the Binet scale in America. He translated Binet's articles into English, applied his tests, and agitated for their general use. He agreed with Binet that the tests worked best in identifying people just below the normal range—Goddard's newly christened morons. But the resemblance between Binet and Goddard ends there. Binet refused to define his scores as "intelligence," and wished to identify in order to help. Goddard regarded the scores as measures of a single, innate entity. He wished to identify in order to recognize limits, segregate, and curtail breeding to prevent further deterioration of an endangered American stock, threatened by immigration from without and by prolific reproduction of its feeble-minded within.

A UNILINEAR SCALE OF INTELLIGENCE

The attempt to establish a unilinear classification of mental deficiency, a rising scale from idiots to imbeciles to morons, embodies two common fallacies pervading most theories of biological determinism discussed in this book: the reification of intelligence as a single, measurable entity; and the assumption, extending back to Morton's skulls (pp. 50–69) and forward to Jensen's universal scaling of general intelligence (pp. 317–320), that evolution is a tale of unilinear progress, and that a single scale ascending from primitive to advanced represents the best way of ordering variation. The concept of progress is a deep prejudice with an ancient pedigree (Bury, 1920) and a subtle power, even over those who would deny it explicitly (Nisbet, 1980).

Can the plethora of causes and phenomena grouped under the rubric of mental deficiency possibly be ordered usefully on a single scale, with its implication that each person owes his rank to the relative amount of a single substance—and that mental deficiency means having less than most? Consider some phenomena mixed up in the common numbers once assigned to defectives of high grade: general low-level mental retardation, specific learning disabilities caused by local neurological damage, environmental disadvantages, cultural differences, hostility to testers. Consider some of the potential causes: inherited patterns of function, genetic pathologies arising accidentally and not passed in family lines, congenital brain damage caused by maternal illness during pregnancy, birth traumas, poor nutrition of fetuses and babies, a variety of environmental disadvantages in early and later life. Yet, to Goddard, all people with mental ages between eight and twelve were morons, all to be treated in roughly the same way: institutionalized or carefully regulated, made happy by catering to their limits, and, above all, prevented from breeding.

Goddard may have been the most unsubtle hereditarian of all. He used his unilinear scale of mental deficiency to identify intelligence as a single entity, and he assumed that everything important about it was inborn and inherited in family lines. He wrote in 1920 (quoted in Tuddenham, 1962, p. 491):

> Stated in its boldest form, our thesis is that the chief determiner of human conduct is a unitary mental process which we call intelligence: that this process is conditioned by a nervous mechanism which is inborn: that the degree of efficiency to be attained by that nervous mechanism and the consequent grade of intellectual or mental level for each individual is determined by the kind of chromosomes that come together with the union of the germ cells: that it is but little affected by any later influences except such serious accidents as may destroy part of the mechanism.

Goddard extended the range of social phenomena caused by differences in innate intelligence until it encompassed almost everything that concerns us about human behavior. Beginning with morons, and working up the scale, he attributed most undesirable behavior to inherited mental deficiency of the offenders. Their problems are caused not only by stupidity per se, but by the link between deficient intelligence and immorality.* High intelligence not only permits us to do our sums; it also engenders the good judgment that underlies all moral behavior.

> The intelligence controls the emotions and the emotions are controlled in proportion to the degree of intelligence. . . . It follows that if there is

*The link of morality to intelligence was a favorite eugenical theme. Thorndike (1940, pp. 264–265), refuting a popular impression that all monarchs are reprobates, cited a correlation coefficient of 0.56 for the estimated intelligence vs. the estimated morality of 269 male members of European royal families!

little intelligence the emotions will be uncontrolled and whether they be strong or weak will result in actions that are unregulated, uncontrolled and, as experience proves, usually undesirable. Therefore, when we measure the intelligence of an individual and learn that he has so much less than normal as to come within the group that we call feeble-minded, we have ascertained by far the most important fact about him (1919, p. 272).

Many criminals, most alcoholics and prostitutes, and even the "ne'er do wells" who simply don't fit in, are morons: "We know what feeble-mindedness is, and we have come to suspect all persons who are incapable of adapting themselves to their environment and living up to the conventions of society or acting sensibly, of being feeble-minded" (1914, p. 571).

At the next level of the merely dull, we find the toiling masses, doing what comes naturally. "The people who are doing the drudgery," Goddard writes (1919, p. 246), "are, as a rule, in their proper places."

> We must next learn that there are great groups of men, laborers, who are but little above the child, who must be told what to do and shown how to do it; and who, if we would avoid disaster, must not be put into positions where they will have to act upon their own initiative or their own judgment. . . . There are only a few leaders, most must be followers (1919, pp. 243–244).

At the upper end, intelligent men rule in comfort and by right. Speaking before a group of Princeton undergraduates in 1919, Goddard proclaimed:

> Now the fact is, that workmen may have a 10 year intelligence while you have a 20. To demand for him such a home as you enjoy is as absurd as it would be to insist that every laborer should receive a graduate fellowship. How can there be such a thing as social equality with this wide range of mental capacity?

"Democracy," Goddard argued (1919, p. 237), "means that the people rule by selecting the wisest, most intelligent and most human to tell them what to do to be happy. Thus Democracy is a method for arriving at a truly benevolent aristocracy."

BREAKING THE SCALE INTO MENDELIAN COMPARTMENTS

But if intelligence forms a single and unbroken scale, how can we solve the social problems that beset us? For at one level, low intelligence generates sociopaths, while at the next grade, industrial society needs docile and dull workers to run its machinery and accept low recompence. How can we convert the unbroken scale into two categories at this crucial point, and still maintain the idea that intelligence is a single, inherited entity? We can now understand why Goddard lavished so much attention upon the moron. The moron threatens racial health because he ranks highest among the undesirable and might, if not identified, be allowed to flourish and propagate. We all recognize the idiot and imbecile and know what must be done; the scale must be broken just above the level of the moron.

> The idiot is not our greatest problem. He is indeed loathsome. . . . Nevertheless, he lives his life and is done. He does not continue the race with a line of children like himself. . . . It is the moron type that makes for us our great problem (1912, pp. 101–102).

Goddard worked in the first flourish of excitement that greeted the rediscovery of Mendel's work and the basic deciphering of heredity. We now know that virtually every major feature of our body is built by the interaction of many genes with each other and with an external environment. But in these early days, many biologists naively assumed that all human traits would behave like the color, size, or wrinkling of Mendel's peas: they believed, in short, that even the most complex parts of a body might be built by single genes, and that variation in anatomy or behavior would record the different dominant and recessive forms of these genes. Eugenicists seized upon this foolish notion with avidity, for it allowed them to assert that all undesirable traits might be traced to single genes and eliminated with proper strictures upon breeding. The early literature of eugenics is filled with speculations, and pedigrees laboriously compiled and fudged, about the gene for *Wanderlust* traced through the family lines of naval captains, or *the* gene for temperament that makes some of us placid and others domineering. We must not be misled by how silly such ideas seem today; they represented orthodox genetics for a brief time, and had a major social impact in America.

Goddard joined the transient bandwagon with a hypothesis that must represent an ultimate in the attempted reification of intelligence. He tried to trace the pedigrees of mental defectives in his Vineland School and concluded that "feeble-mindedness" obeyed Mendelian rules of inheritance. Mental deficiency must therefore be a definite thing, and it must be governed by a single gene, undoubtedly recessive to normal intelligence (1914, p. 539). "Normal intelligence," Goddard concluded, "seems to be a unit character and transmitted in true Mendelian fashion" (1914, p. ix).

Goddard claimed that he had been compelled to make this unlikely conclusion by the press of evidence, not by any prior hope or prejudice.

> Any theories or hypotheses that have been presented have been merely those that were suggested by the data themselves, and have been worked out in an effort to understand what the data seem to comprise. Some of the conclusions are as surprising to the writer and as difficult for him to accept as they are likely to be to many readers (1914, p. viii).

Can we seriously view Goddard as a forced and reluctant convert to a hypothesis that fit his general scheme so well and solved his most pressing problem so neatly? A single gene for normal intelligence removed the potential contradiction between a unilinear scale that marked intelligence as a single, measurable entity, and a desire to separate and identify the mentally deficient as a category apart. Goddard had broken his scale into two sections at just the right place: morons carried a double dose of the bad recessive; dull laborers had at least one copy of the normal gene and could be set before their machines. Moreover, the scourge of feeble-mindedness might now be eliminated by schemes of breeding easily planned. One gene can be traced, located, and bred out. If one hundred genes regulate intelligence, eugenic breeding must fail or proceed with hopeless sloth.

THE PROPER CARE AND FEEDING (BUT NOT BREEDING) OF MORONS

If mental deficiency is the effect of a single gene, the path to its eventual elimination lies evidently before us: do not allow such people to bear children:

> If both parents are feeble-minded all the children will be feeble-minded. It is obvious that such matings should not be allowed. It is perfectly clear that no feeble-minded person should ever be allowed to marry or to become a parent. It is obvious that if this rule is to be carried out the intelligent part of society must enforce it (1914, p. 561).

If morons could control their own sexual urges and desist for the good of mankind, we might permit them to live freely among us. But they cannot, because immorality and stupidity are inexorably linked. The wise man can control his sexuality in a rational manner: "Consider for a moment the sex emotion, supposed to be the most uncontrollable of all human instincts; yet it is notorious that the intelligent man controls even this" (1919, p. 273). The moron cannot behave in so exemplary and abstemious a fashion:

> They are not only lacking in control but they are lacking often in the perception of moral qualities; if they are not allowed to marry they are nevertheless not hindered from becoming parents. So that if we are absolutely to prevent a feeble-minded person from becoming a parent, something must be done other than merely prohibiting the marrying. To this end there are two proposals: the first is colonization, the second is sterilization (1914, p. 566).

Goddard did not oppose sterilization, but he regarded it as impractical because traditional sensibilities of a society not yet wholly rational would prevent such widespread mayhem. Colonization in exemplary institutions like his own at Vineland, New Jersey, must be our preferred solution. Only here could the reproduction of morons be curtailed. If the public balked at the great expense of building so many new centers for confinement, the cost could easily be recouped by its own savings:

> If such colonies were provided in sufficient number to take care of all the distinctly feeble-minded cases in the community, they would very largely take the place of our present almshouses and prisons, and they would greatly decrease the numbers in our insane hospitals. Such colonies would save an annual loss in property and life, due to the action of these irresponsible people, sufficient to nearly, or quite, offset the expense of the new plant (1912, pp. 105–106).

Inside these institutions, morons could operate in contentment at their biologically appointed level, denied only the basic biology of their own sexuality. Goddard ended his book on the causes of mental deficiency with this plea for the care of institutionalized morons: "Treat them as children according to their mental age, constantly encourage and praise, never discourage or scold; and *keep them happy*" (1919, p. 327).

Preventing the immigration and propagation of morons

Once Goddard had identified the cause of feeble-mindedness in a single gene, the cure seemed simple enough: don't allow native morons to breed and keep foreign ones out. As a contribution to the second step, Goddard and his associates visited Ellis Island in 1912 "to observe conditions and offer any suggestions as to what might be done to secure a more thorough examination of immigrants for the purpose of detecting mental defectives" (Goddard, 1917, p. 253).

As Goddard described the scene, a fog hung over New York harbor that day and no immigrants could land. But one hundred were about ready to leave, when Goddard intervened: "We picked out one young man whom we suspected was defective, and, through the interpreter, proceeded to give him the test. The boy tested 8 by the Binet scale. The interpreter said, 'I could not have done that when I came to this country,' and seemed to think the test unfair. We convinced him that the boy was defective" (Goddard, 1913, p. 105).

Encouraged by this, one of the first applications of the Binet scale in America, Goddard raised some funds for a more thorough study and, in the spring of 1913, sent two women to Ellis Island for two and a half months. They were instructed to pick out the feeble-minded by sight, a task that Goddard preferred to assign to women, to whom he granted innately superior intuition:

> After a person has had considerable experience in this work, he almost gets a sense of what a feeble-minded person is so that he can tell one afar off. The people who are best at this work, and who I believe should do this work, are women. Women seem to have closer observation than men. It was quite impossible for others to see how these two young women could pick out the feeble-minded without the aid of the Binet test at all (1913, p. 106).

Goddard's women tested thirty-five Jews, twenty-two Hungarians, fifty Italians, and forty-five Russians. These groups could not be regarded as random samples because government officials had already "culled out those they recognized as defective." To balance this bias, Goddard and his associates "passed by the obviously normal. That left us the great mass of 'average immigrants.' " (1917, p. 244). (I am continually amazed by the unconscious statements of prejudice that slip into supposedly objective accounts. Note here that average immigrants are below normal, or at least not obviously normal—the proposition that Goddard was supposedly testing, not asserting a priori.)

Binet tests on the four groups led to an astounding result: 83 percent of the Jews, 80 percent of the Hungarians, 79 percent of the Italians, and 87 percent of the Russians were feeble-minded—that is, below age twelve on the Binet scale. Goddard himself was flabbergasted: could anyone be made to believe that four-fifths of any nation were morons? "The results obtained by the foregoing evaluation of the data are so surprising and difficult of acceptance that they can hardly stand by themselves as valid" (1917, p. 247). Perhaps the tests had not been adequately explained by interpreters? But the Jews had been tested by a Yiddish-speaking psychologist, and they ranked no higher than the other groups. Eventually, Goddard monkied about with the tests, tossed several out, and got his figures down to 40 to 50 percent, but still he was disturbed.

Goddard's figures were even more absurd than he imagined for two reasons, one obvious, the other less so. As a nonevident reason, Goddard's original translation of the Binet scale scored people harshly and made morons out of subjects usually regarded as normal. When Terman devised the Stanford-Binet scale in 1916, he found that Goddard's version ranked people well below his own. Terman reports (1916, p. 62) that of 104 adults tested by him as between twelve and fourteen years mental age (low, but normal intelligence), 50 percent were morons on the Goddard scale.

For the evident reason, consider a group of frightened men and women who speak no English and who have just endured an oceanic voyage in steerage. Most are poor and have never gone to school; many have never held a pencil or pen in their hand. They march off the boat; one of Goddard's intuitive women takes them aside shortly thereafter, sits them down, hands them a pencil, and asks them to reproduce on paper a figure shown to them a moment ago, but now withdrawn from their sight. Could their failure be a result of testing conditions, of weakness, fear, or confusion, rather than of innate stupidity? Goddard considered the possibility, but rejected it:

> The next question is 'drawing a design from memory,' which is passed by only 50 percent. To the uninitiated this will not seem surprising since it looks hard, and even those who are familiar with the fact that normal children of 10 pass it without difficulty may admit that persons who have never had a pen or pencil in their hands, as was true of many of the immigrants, may find it impossible to draw the design (1917, p. 250).

Permitting a charitable view of this failure, what but stupidity could explain an inability to state more than sixty words, any words, in one's own language during three minutes?

> What shall we say of the fact that only 45 percent can give 60 words in three minutes, when normal children of 11 years sometimes give 200 words in that time! It is hard to find an explanation except lack of intelligence or lack of vocabulary, and such a lack of vocabulary in an adult would probably mean lack of intelligence. How could a person live even 15 years in any environment without learning hundreds of names of which he could certainly think of 60 in three minutes? (1917, p. 251)

Or ignorance of the date, or even the month or year?

> Must we again conclude that the European peasant of the type that immigrates to America pays no attention to the passage of time? That the drudgery of life is so severe that he cares not whether it is January or July, whether it is 1912 or 1906? Is it possible that the person may be of considerable intelligence and yet, because of the peculiarity of his environment, not have acquired this ordinary bit of knowledge, even though the calendar is not in general use on the continent, or is somewhat complicated as in Russia? If so what an environment it must have been! (1917, p. 250)

Since environment, either European or immediate, could not explain such abject failure, Goddard stated: "We cannot escape the general conclusion that these immigrants were of surprisingly low intelligence" (1917, p. 251). The high proportion of morons still bothered Goddard, but he finally attributed it to the changing character of immigration: "It should be noted that the immigration of recent years is of a decidedly different character from the early immigration. . . . We are now getting the poorest of each race" (1917, p. 266). "The intelligence of the average 'third class' immigrant is low, perhaps of moron grade" (1917, p. 243). Perhaps, Goddard hoped out loud, things were better on the upper decks, but he did not test these wealthier customers.

What then should be done? Should all these morons be shipped back, or prevented from starting out in the first place? Foreshadowing the restrictions that would be legislated within a decade, Goddard argued that his conclusions "furnish important considerations for future actions both scientific and social as well as legislative" (1917, p. 261). But by this time Goddard had softened his earlier harsh position on the colonization of morons. Perhaps there were not enough merely dull workers to fill the vast number of frankly undesirable jobs. The moron might have to be recruited: "They do a great deal of work that no one else will do. . . . There is an immense amount of drudgery to be done, an immense amount of work for which we do not wish to pay enough to secure more intelligent workers. . . . May it be that possibly the moron has his place" (1917, p. 269).

Nonetheless, Goddard rejoiced in the general tightening of standards for admission. He reports that deportations for mental deficiency increased 350 percent in 1913 and 570 percent in 1914 over the average of the five preceding years:

> This was due to the untiring efforts of the physicians who were inspired by the belief that mental tests could be used for the detection of feeble-minded aliens. . . . If the American public wishes feeble-minded aliens excluded, it must demand that congress provide the necessary facilities at the ports of entry (1917, p. 271).

Meanwhile, at home, the feeble-minded must be identified and kept from breeding. In several studies, Goddard exposed the menace of moronity by publishing pedigrees of hundreds of worthless souls, charges upon the state and community, who would never have been born had their feeble-minded forebears been debarred from reproduction. Goddard discovered a stock of paupers and ne'er-do-wells in the pine barrens of New Jersey and traced their ancestry back to the illicit union of an upstanding man with a supposedly feeble-minded tavern wench. The same man later married a worthy Quakeress and started another line composed wholly of upstanding citizens. Since the progenitor had fathered both a good and a bad line, Goddard combined the Greek words for beauty (*kallos*) and bad (*kakos*), and awarded him the pseudonym Martin Kallikak. Goddard's Kallikak family functioned as a primal myth of the eugenics movement for several decades.

Goddard's study is little more than guesswork rooted in conclusions set from the start. His method, as always, rested upon the training of intuitive women to recognize the feeble-minded by sight. Goddard did not administer Binet tests in pine-barren shacks. Goddard's faith in visual identification was virtually unbounded. In 1919 he analyzed Edwin Markham's poem "The Man With The Hoe":

> Bowed by the weight of centuries he leans
> Upon his hoe and gazes at the ground,
> The emptiness of ages in his face
> And on his back the burden of the world. . . .

Markham's poem had been inspired by Millet's famous painting of the same name. The poem, Goddard complained (1919, p. 239), "seems to imply that the man Millet painted came to his condition as the result of social conditions which held him down and made him like the clods that he turned over." Nonsense, exclaimed Goddard; most poor peasants suffer only from their own feeble-mindedness, and Millet's painting proves it. Couldn't Markham see that the peasant is mentally deficient? "Millet's Man With The Hoe is a man of arrested mental development—the painting is a perfect picture of an imbecile" (1919, pp. 239–240). To Markham's searing question: "Whose breath blew out the light within this brain," Goddard replied that mental fire had never been kindled.

Since Goddard could determine degrees of mental deficiency by examining a painting, he certainly anticipated no trouble with flesh and blood. He dispatched the redoubtable Ms. Kite, soon to see further service on Ellis Island, to the pine barrens and quickly produced the sad pedigree of the kakos line. Goddard describes one of Ms. Kite's identifications (1912, pp. 77–78):

> Used as she was to the sights of misery and degradation, she was hardly prepared for the spectacle within. The father, a strong, healthy, broad-shouldered man, was sitting helplessly in a corner. . . . Three children, scantily clad and with shoes that would barely hold together, stood about drooping jaws and the unmistakable look of the feeble-minded. . . . The whole family was a living demonstration of the futility of trying to make desirable citizens from defective stock through making and enforcing compulsory education laws. . . . The father himself, though strong and vigorous, showed by his face that he had only a child's mentality. The mother in her filth and rags was also a child. In this house of abject poverty, only one sure prospect was ahead, that it would produce more feeble-minded children with which to clog the wheels of human progress.

If these spot identifications seem a bit hasty or dubious, consider Goddard's method for inferring the mental state of the departed, or otherwise unavailable (1912, p. 15):

5•1 An honest picture of Deborah, the Kallikak descendant living in Goddard's institution.

> After some experience, the field worker becomes expert in inferring the condition of those persons who are not seen, from the similarity of the language used in describing them to that used in describing persons she has seen.

It may be a small item in the midst of such absurdity, but I discovered a bit of more conscious skulduggery two years ago. My colleague Steven Selden and I were examining his copy of Goddard's volume of the Kallikaks. The frontispiece shows a member of the kakos line, saved from depravity by confinement in Goddard's institution at Vineland. Deborah, as Goddard calls her, is a beautiful woman (Fig. 5.1). She sits calmly in a white dress, reading a book, a cat lying comfortably on her lap. Three other plates show members of the kakos line, living in poverty in their rural shacks. All have a depraved look about them (Fig. 5.2). Their mouths are sinister in appearance; their eyes are darkened slits. But Goddard's books are nearly seventy years old, and the ink has faded. It is now clear that all the photos of noninstitutionalized kakos were phonied by inserting heavy dark lines to give eyes and mouths their diabolical appearance. The three plates of Deborah are unaltered.

Selden took his book to Mr. James H. Wallace, Jr., director of Photographic Services at the Smithsonian Institution. Mr. Wallace reports (letter to Selden, 17 March 1980):

> There can be no doubt that the photographs of the Kallikak family members have been retouched. Further, it appears that this retouching was limited to the facial features of the individuals involved—specifically eyes, eyebrows, mouths, nose and hair.
>
> By contemporary standards, this retouching is extremely crude and obvious. It should be remembered, however, that at the time of the original publication of the book, our society was far less visually sophisticated. The widespread use of photographs was limited, and casual viewers of the time would not have nearly the comparative ability possessed by even preteenage children today. . . .
>
> The harshness clearly gives the appearance of dark, staring features, sometimes evilness, and sometimes mental retardation. It would be difficult to understand why any of this retouching was done were it not to give the viewer a false impression of the characteristics of those depicted. I believe the fact that no other areas of the photographs, or the individuals have been retouched is significant in this regard also. . . .
>
> I find these photographs to be an extremely interesting variety of photographic manipulation.

Goddard recants

By 1928 Goddard had changed his mind and become a latter-day supporter of the man whose work he had originally perverted, Alfred Binet. Goddard admitted, first of all, that he had set the upper limit of moronity far too high:

> It was for a time rather carelessly assumed that everybody who tested 12 years or less was feeble-minded. . . . We now know, of course, that only a small percentage of the people who test 12 are actually feeble-minded—that is, are incapable of managing their affairs with ordinary prudence or of competing in the struggle for existence (1928, p. 220).

But genuine morons still abound at their redefined level. What shall be done with them? Goddard did not abandon his belief in their inherited mentality, but he now took Binet's line and argued that most, if not all, could be trained to lead useful lives in society:

> The problem of the moron is a problem of education and training. . . . This may surprise you, but frankly when I see what has been made out of the moron by a system of education, which as a rule is only half right, I have no difficulty in concluding that when we get an education that is entirely right there will be no morons who cannot manage themselves and their affairs and compete in the struggle for existence. If we could hope to add to this a social order that would literally give every man a chance, I should be perfectly sure of the result (1928, pp. 223–224).

But if we let morons live in society, will they not marry and bear children; is this not the greatest danger of all, the source of Goddard's previous and passionate warnings?

> Some will object that this plan neglects the eugenic aspect of the problem. In the community, these morons will marry and have children. And why not? . . . It may still be objected that moron parents are likely to have imbecile or idiot children. There is not much evidence that this is the case. The danger is probably negligible. At least it is not likely to occur any

---

5•2 Altered photographs of members of the Kallikak family living in poverty in the New Jersey pine barrens. Note how mouths and eyebrows are accentuated to produce an appearance of evil or stupidity. The effect is much clearer on the original photographs produced in Goddard's book.

oftener than it does in the general population.* I assume that most of you, like myself, will find it difficult to admit that the foregoing may be the true view. We have worked too long under the old concept (1928, pp. 223–224).

Goddard concluded (1928, p. 225) in reversing the two bulwarks of his former system:

1. Feeble-mindedness (the moron) is *not incurable* [Goddard's italics].

2. The feeble-minded do not generally need to be segregated in institutions.

"As for myself," Goddard confessed (p. 224), "I think I have gone over to the enemy."

Lewis M. Terman and the mass marketing of innate IQ

Without offering any data on all that occurs between conception and the age of kindergarten, they announce on the basis of what they have got out of a few thousand questionnaires that they are measuring the hereditary mental endowment of human beings. Obviously, this is not a conclusion obtained by research. It is a conclusion planted by the will to believe. It is, I think, for the most part unconsciously planted. . . . If the impression takes root that these tests really measure intelligence, that they constitute a sort of last judgment on the child's capacity, that they reveal "scientifically" his predestined ability, then it would be a thousand times better if all the intelligence testers and all their questionnaires were sunk without warning in the Sargasso Sea.

—WALTER LIPPMANN, in the course of a debate with Lewis Terman

Mass testing and the Stanford-Binet

Lewis M. Terman, the twelfth child in an Indiana farm family of fourteen, traced his interest in the study of intelligence to an itinerant book peddler and phrenologist who visited his home when he was nine or ten and predicted good things after feeling the bumps on his skull. Terman pursued this early interest, never doubting that a measurable mental worth lay inside people's heads. In his doctoral dissertation of 1906, Terman examined seven "bright" and seven "stupid" boys and defended each of his tests as a measure of intelligence by appealing to the standard catalogue of

*Do not read into this statement more than Goddard intended. He had not abandoned his belief in the heritability of moronity itself. Moron parents will have moron children, but they can be made useful through education. Moron parents, however, do not preferentially beget defectives of *lower* grade—idiots and imbeciles.

racial and national stereotypes. Of tests for invention, he wrote: "We have only to compare the negro with the Eskimo or Indian, and the Australian native with the Anglo-Saxon, to be struck by an apparent kinship between general intellectual and inventive ability" (1906, p. 14). Of mathematical ability, he proclaimed (1906, p. 29): "Ethnology shows that racial progress has been closely paralleled by development of the ability to deal with mathematical concepts and relations."

Terman concluded his study by committing both of the fallacies identified on p. 155 as foundations of the hereditarian view. He reified average test scores as a "thing" called general intelligence by advocating the first of two possible positions (1906, p. 9): "Is intellectual ability a bank account, on which we can draw for any desired purpose, or is it rather a bundle of separate drafts, each drawn for a specific purpose and inconvertible?" And, while admitting that he could provide no real support for it, he defended the innatist view (1906, p. 68): "While offering little positive data on the subject, the study has strengthened my impression of the relatively greater importance of endowment over training as a determinant of an individual's intellectual rank among his fellows."

Goddard introduced Binet's scale to America, but Terman was the primary architect of its popularity. Binet's last version of 1911 included fifty-four tasks, graded from prenursery to mid-teen-age years. Terman's first revision of 1916 extended the scale to "superior adults" and increased the number of tasks to ninety. Terman, by then a professor at Stanford University, gave his revision a name that has become part of our century's vocabulary—the Stanford-Binet, the standard for virtually all "IQ" tests that followed.*

I offer no detailed analysis of content (see Block and Dworkin, or Chase, 1977), but present two examples to show how Terman's tests stressed conformity with expectation and downgraded original response. When expectations are society's norms, then do the

*Terman (1919) provided a lengthy list of the attributes of general intelligence captured by the Stanford-Binet tests: memory, language comprehension, size of vocabulary, orientation in space and time, eye-hand coordination, knowledge of familiar things, judgment, likeness and differences, arithmetical reasoning, resourcefulness and ingenuity in difficult practical situations, ability to detect absurdities, speed and richness of association of ideas, power to combine the dissected parts of a form board or a group of ideas into a unitary whole, capacity to generalize from particulars, and ability to deduce a rule from connected facts.

tests measure some abstract property of reasoning, or familiarity with conventional behavior? Terman added the following item to Binet's list:

> An Indian who had come to town for the first time in his life saw a white man riding along the street. As the white man rode by, the Indian said—'The white man is lazy; he walks sitting down.' What was the white man riding on that caused the Indian to say, 'He walks sitting down.'

Terman accepted "bicycle" as the only correct response—not cars or other vehicles because legs don't go up and down in them; not horses (the most common "incorrect" answer) because any self-respecting Indian would have known what he was looking at. (I myself answered "horse," because I saw the Indian as a clever ironist, criticizing an effete city relative.) Such original responses as "a cripple in a wheel chair," and "a person riding on someone's back" were also marked wrong.

Terman also included this item from Binet's original: "My neighbor has been having queer visitors. First a doctor came to his house, then a lawyer, then a minister. What do you think happened there?" Terman permitted little latitude beyond "a death," though he did allow "a marriage" from a boy he described as "an enlightened young eugenist" who replied that the doctor came to see if the partners were fit, the lawyer to arrange, and the minister to tie the knot. He did not accept the combination "divorce and remarriage," though he reports that a colleague in Reno, Nevada, had found the response "very, very common." He also did not permit plausible but uncomplicated solutions (a dinner, or an entertainment), or such original responses as: "someone is dying and is getting married and making his will before he dies."

But Terman's major influence did not reside in his sharpening or extension of the Binet scale. Binet's tasks had to be administered by a trained tester working with one child at a time. They could not be used as instruments for general ranking. But Terman wished to test everybody, for he hoped to establish a gradation of innate ability that could sort all children into their proper stations in life:

> What pupils shall be tested? The answer is, all. If only selected children are tested, many of the cases most in need of adjustment will be over-

looked. The purpose of the tests is to tell us what we do not already know, and it would be a mistake to test only those pupils who are recognized as obviously below or above average. Some of the biggest surprises are encountered in testing those who have been looked upon as close to average in ability. Universal testing is fully warranted (1923, p. 22).

The Stanford-Binet, like its parent, remained a test for individuals, but it became the paradigm for virtually all the written versions that followed. By careful juggling and elimination,* Terman standardized the scale so that "average" children would score 100 at each age (mental age equal to chronological age). Terman also evened out the variation among children by establishing a standard deviation of 15 or 16 points at each chronological age. With its mean of 100 and standard deviation of 15, the Stanford-Binet became (and in many respects remains to this day) the primary criterion for judging a plethora of mass-marketed written tests that followed. The invalid argument runs: we know that the Stanford-Binet measures intelligence; therefore, any written test that correlates strongly with Stanford-Binet also measures intelligence. Much of the elaborate statistical work performed by testers during the past fifty years provides no independent confirmation for the proposition that tests measure intelligence, but merely establishes correlation with a preconceived and unquestioned standard.

Testing soon became a multimillion-dollar industry; marketing companies dared not take a chance with tests not proven by their correlation with Terman's standard. The Army Alpha (see pp. 192–222) initiated mass testing, but a flood of competitors greeted school administrators within a few years after the war's end. A quick glance at the advertisements appended to Terman's later book (1923) illustrates, dramatically and unintentionally, how all Terman's cautious words about careful and lengthy assessment (1919, p. 299, for example) could evaporate before strictures of cost and time when his desire to test all children became a reality (Fig. 5.3). Thirty minutes and five tests might mark a child for life, if schools adopted the following examination, advertised in Terman 1923, and constructed by a committee that included Thorndike, Yerkes, and Terman himself.

*This, in itself, is not finagling, but a valid statistical procedure for establishing uniformity of average score and variance across age levels.

Prepared under the auspices of the National Research Council

NATIONAL INTELLIGENCE TESTS

By M. E. HAGGERTY, L. M. TERMAN, E. L. THORNDIKE
G. M. WHIPPLE, and R. M. YERKES

THESE tests are the direct result of the application of the army testing methods to school needs. They were devised in order to supply group tests for the examination of school children that would embody the greater benefits derived from the Binet and similar tests.

The effectiveness of the army intelligence tests in problems of classification and diagnosis is a measure of the success that may be expected to attend the use of the National Intelligence Tests, which have been greatly improved in the light of army experiences.

The tests have been selected from a large group of tests after a try-out and a careful analysis by a statistical staff. The two scales prepared consist of five tests each (with practice exercises), and either may be administered in thirty minutes. They are simple in application, reliable, and immediately useful for classifying children in Grades 3 to 8 with respect to intellectual ability. Scoring is unusually simple.

Either scale may be used separately to advantage. The reliability of results is increased, however, by reexamination with the other scale after an interval of at least a day.

Scale A consists of an arithmetical reasoning, a sentence completion, a logical selection, a synonym-antonym, and a symbol-digit test. Scale B includes a completion, an information, a vocabulary, an analogies, and a comparison test.

Scale A: *Form* 1. 12 pages. Price per package of 25 Examination Booklets, 2 Scoring Keys, and 1 Class Record $1.45 net.
Scale A: *Form* 2. Same description. Same price.
Scale B: *Form* 1. 12 pages. Price per package of 25 Examination Booklets, Scoring Key, and Class Record $1.45 net.
Scale B: *Form* 2. Same description. Same price.
Manual of Directions. Paper. 32 pages. Price 25 cents net.
Specimen Set. One copy of each Scale and Scoring Keys and Manual of Directions. Price 50 cents postpaid.

***Experimental work financed by the General Education Board by appropriation of* $25,000**

WORLD BOOK COMPANY

YONKERS-ON-HUDSON, NEW YORK
2126 PRAIRIE AVENUE, CHICAGO

5•3 An advertisement for mass mental testing using an examination written by, among others, Terman and Yerkes.

> *National Intelligence Tests for Grades 3–8*
>
> The direct result of the application of the army testing methods to school needs. . . . The tests have been selected from a large group of tests after a try-out and a careful analysis by a statistical staff. The two scales prepared consist of five tests each (with practical exercises) and either may be administered in thirty minutes. They are simple in application, reliable, and immediately useful in classifying children in Grades 3 to 8 with respect to intellectual ability. Scoring is unusually simple.

Binet, had he lived, might have been distressed enough by such a superficial assessment, but he would have reacted even more strongly against Terman's intent. Terman agreed with Binet that the tests worked best for identifying "high-grade defectives," but his reasons for so doing stand in chilling contrast with Binet's desire to segregate and help (1916, pp. 6–7):

> It is safe to predict that in the near future intelligence tests will bring tens of thousands of these high-grade defectives under the surveillance and protection of society. This will ultimately result in curtailing the reproduction of feeble-mindedness and in the elimination of an enormous amount of crime, pauperism, and industrial inefficiency. It is hardly necessary to emphasize that the high-grade cases, of the type now so frequently overlooked, are precisely the ones whose guardianship it is most important for the State to assume.

Terman relentlessly emphasized limits and their inevitability. He needed less than an hour to crush the hopes and belittle the efforts of struggling, "well-educated" parents afflicted with a child of IQ 75.

> Strange to say, the mother is encouraged and hopeful because she sees that her boy is learning to read. She does not seem to realize that at his age he ought to be within three years of entering high school. The forty-minute test has told more about the mental ability of this boy than the intelligent mother had been able to learn in eleven years of daily and hourly observation. For X is feeble-minded; he will never complete the grammar school; he will never be an efficient worker or a responsible citizen (1916).

Walter Lippmann, then a young journalist, saw through Terman's numbers to the heart of his preconceived attempt, and wrote in measured anger:

> The danger of the intelligence tests is that in a wholesale system of education, the less sophisticated or the more prejudiced will stop when

they have classified and forget that their duty is to educate. They will grade the retarded child instead of fighting the causes of his backwardness. For the whole drift of the propaganda based on intelligence testing is to treat people with low intelligence quotients as congenitally and hopelessly inferior.

Terman's technocracy of innateness

If it were true, the emotional and worldly satisfactions in store for the intelligence tester would be very great. If he were really measuring intelligence, and if intelligence were a fixed hereditary quantity, it would be for him to say not only where to place each child in school, but also which children should go to high school, which to college, which into the professions, which into the manual trades and common labor. If the tester would make good his claim, he would soon occupy a position of power which no intellectual has held since the collapse of theocracy. The vista is enchanting, and even a little of the vista is intoxicating enough. If only it could be proved, or at least believed, that intelligence is fixed by heredity, and that the tester can measure it, what a future to dream about! The unconscious temptation is too strong for the ordinary critical defenses of the scientific methods. With the help of a subtle statistical illusion, intricate logical fallacies and a few smuggled *obiter dicta*, self-deception as the preliminary to public deception is almost automatic. —WALTER LIPPMANN, in a debate with Terman

Plato had dreamed of a rational world ruled by philosopher-kings. Terman revived this dangerous vision but led his corps of mental testers in an act of usurpation. If all people could be tested, and then sorted into roles appropriate for their intelligence, then a just, and, above all, efficient society might be constructed for the first time in history.

Dealing off the bottom, Terman argued that we must first restrain or eliminate those whose intelligence is too low for an effective or moral life. The primary cause of social pathology is innate feeble-mindedness. Terman (1916, p. 7) criticized Lombroso for thinking that the externalities of anatomy might record criminal behavior. Innateness, to be sure, is the source, but its direct sign is low IQ, not long arms or a jutting jaw:

The theories of Lombroso have been wholly discredited by the results of intelligence tests. Such tests have demonstrated, beyond any possibility of doubt, that the most important trait of at least 25 percent of our criminals is mental weakness. The physical abnormalities which have been found so common among prisoners are not the stigmata of criminality, but the physical accompaniments of feeble-mindedness. They have no diagnostic significance except in so far as they are indications of mental deficiency (1916, p. 7).

Feeble-minded people are doubly burdened by their unfortunate inheritance, for lack of intelligence, debilitating enough in itself, leads to immorality. If we would eliminate social pathology, we must identify its cause in the biology of sociopaths themselves—and then eliminate them by confinement in institutions and, above all, by preventing their marriage and the production of offspring.

> Not all criminals are feeble-minded, but all feeble-minded persons are at least potential criminals. That every feeble-minded woman is a potential prostitute would hardly be disputed by anyone. Moral judgment, like business judgment, social judgment, or any other kind of higher thought process, is a function of intelligence. Morality cannot flower and fruit if intelligence remains infantile (1916, p.11).
>
> The feeble-minded, in the sense of social incompetents, are by definition a burden rather than an asset, not only economically but still more because of their tendencies to become delinquent or criminal. . . . The only effective way to deal with the hopelessly feeble-minded is by permanent custodial care. The obligations of the public school rest rather with the large and more hopeful group of children who are merely inferior (1919, pp. 132–133).

In a plea for universal testing, Terman wrote (1916, p. 12): "Considering the tremendous cost of vice and crime, which in all probability amounts to not less than $500,000,000 per year in the United States alone, it is evident that psychological testing has found here one of its richest applications."

After marking the sociopath for removal from society, intelligence tests might then channel biologically acceptable people into professions suited for their mental level. Terman hoped that his testers would "determine the minimum 'intelligence quotient' necessary for success in each leading occupation" (1916, p. 17). Any conscientious professor tries to find jobs for his students, but few are audacious enough to tout their disciples as apostles of a new social order:

> Industrial concerns doubtless suffer enormous losses from the employment of persons whose mental ability is not equal to the tasks they are expected to perform. . . . Any business employing as many as 500 or 1000 workers, as, for example, a large department store, could save in this way several times the salary of a well-trained psychologist.

Terman virtually closed professions of prestige and monetary reward to people with IQ below 100 (1919, p. 282), and argued that "substantial success" probably required an IQ above 115 or 120. But he was more interested in establishing ranks at the low end of the scale, among those he had deemed "merely inferior." Modern industrial society needs its technological equivalent of the Biblical metaphor for more bucolic times—the hewers of wood and drawers of water. And there are so many of them:

> The evolution of modern industrial organization together with the mechanization of processes by machinery is making possible the larger and larger utilization of inferior mentality. One man with ability to think and plan guides the labor of ten or twenty laborers, who do what they are told to do and have little need for resourcefulness or initiative (1919, p. 276).

IQ of 75 or below should be the realm of unskilled labor, 75 to 85 "preeminently the range for semi-skilled labor." More specific judgments could also be made. "Anything above 85 IQ in the case of a barber probably represents so much dead waste" (1919, p. 288). IQ 75 is an "unsafe risk in a motorman or conductor, and it conduces to discontent" (Terman, 1919). Proper vocational training and placement is essential for those "of the 70 to 85 class." Without it, they tend to leave school "and drift easily into the ranks of the anti-social or join the army of Bolshevik discontents" (1919, p. 285).

Terman investigated IQ among professions and concluded with satisfaction that an imperfect allocation by intelligence had already occurred naturally. The embarrassing exceptions he explained away. He studied 47 express company employees, for example, men engaged in rote, repetitive work "offering exceedingly limited opportunity for the exercise of ingenuity or even personal judgment" (1919, p. 275). Yet their median IQ stood at 95, and fully 25 percent measured above 104, thus winning a place among the ranks of the intelligent. Terman was puzzled, but attributed such low achievement primarily to a lack of "certain emotional, moral, or other desirable qualities," though he admitted that "economic pressures" might have forced some "out of school before they were able to prepare for more exacting service" (1919, p. 275). In another study, Terman amassed a sample of 256 "hoboes and unemployed," largely from a "hobo hotel" in Palo Alto. He expected to find their average IQ at the bottom of his list; yet, while the hobo mean of 89 did not suggest enormous endowment, they still ranked above motormen, salesgirls, firemen, and policemen. Terman suppressed this embarrassment by ordering his table in a curious way. The hobo mean was distressingly high, but hobos also varied more than any other group, and included a substantial number of rather low scores. So Terman arranged his list by the scores of the lowest 25 percent in each group, and sunk his hobos into the cellar.

Had Terman merely advocated a meritocracy based on achievement, one might still decry his elitism, but applaud a scheme that awarded opportunity to hard work and strong motivation. But Terman believed that class boundaries had been set by innate intelligence. His coordinated rank of professions, prestige, and salaries reflected the biological worth of existing social classes. If barbers did not remain Italian, they would continue to arise from the poor and to stay appropriately among them:

> The common opinion that the child from a cultured home does better in tests solely by reason of his superior home advantages is an entirely gratuitous assumption. Practically all of the investigations which have been made of the influence of nature and nurture on mental performance agree in attributing far more to original endowment than to environment. Common observation would itself suggest that the social class to which the family belongs depends less on chance than on the parents' native qualities of intellect and character. . . . The children of successful and cultured parents test higher than children from wretched and ignorant homes for the simple reason that their heredity is better (1916, p. 115).

Fossil IQ's of past geniuses

Society may need masses of the "merely inferior" to run its machines, Terman believed, but its ultimate health depends upon the leadership of rare geniuses with elevated IQ's. Terman and his associates published a five-volume series on *Genetic Studies of Genius* in an attempt to define and follow people at the upper end of the Stanford-Binet scale.

In one volume, Terman decided to measure, retrospectively, the IQ of history's prime movers—its statesmen, soldiers, and intellectuals. If they ranked at the top, then IQ is surely the single measure of ultimate worth. But how can a fossil IQ be recovered without conjuring up young Copernicus and asking him what the white man was riding? Undaunted, Terman and his colleagues tried to reconstruct the IQ of past notables, and published a thick book (Cox, 1926) that must rank as a primary curiosity within a literature already studded with absurdity—though Jensen (1979, pp. 113 and 355) and others still take it seriously.*

Terman (1917) had already published a preliminary study of Francis Galton and awarded a staggering IQ of 200 to this pioneer of mental testing. He therefore encouraged his associates to proceed with a larger investigation. J. M. Cattell had published a ranking of the 1,000 prime movers of history by measuring the lengths of their entries in biographical dictionaries. Catherine M. Cox, Terman's associate, whittled the list to 282, assembled detailed biographical information about their early life, and proceeded to estimate two IQ values for each—one, called A1 IQ, for birth to seventeen years; the other, A2 IQ, for ages seventeen to twenty-six.

Cox ran into problems right at the start. She asked five people, including Terman, to read her dossiers and to estimate the two IQ scores for each person. Three of the five agreed substantially in their mean values, with A1 IQ clustering around 135 and A2 IQ near 145. But two of the raters differed markedly, one awarding an average IQ well above, the other well below, the common figure. Cox simply eliminated their scores, thereby throwing out 40 percent of her data. Their low and high scores would have balanced each other at the mean in any case, she argued (1926, p. 72). Yet if five people working in the same research group could not agree, what hope for uniformity or consistency—not to mention objectivity—could be offered?

Apart from these debilitating practical difficulties, the basic logic of the study was hopelessly flawed from the first. The differences in IQ that Cox recorded among her subjects do not measure their varying accomplishments, not to mention their native intelligence. Instead, the differences are a methodological artifact of the varying quality of information that Cox was able to compile about the childhood and early youth of her subjects. Cox began by assigning a base IQ of 100 to each individual; the raters then added to (or, rarely, subtracted from) this value according to the data provided.

*Jensen writes: "The average estimated IQ of three hundred historical persons . . . on whom sufficient childhood evidence was available for a reliable estimate was IQ 155. . . . Thus the majority of these eminent men would most likely have been recognized as intellectually gifted in childhood had they been given IQ tests" (Jensen, 1979, p. 113).

Cox's dossiers are motley lists of childhood and youthful accomplishments, with an emphasis on examples of precocity. Since her method involved adding to the base figure of 100 for each notable item in the dossier, estimated IQ records little more than the volume of available information. In general, low IQ's reflect an absence of information, and high IQ's an extensive list. (Cox even admits that she is not measuring true IQ, but only what can be deduced from limited data, though this disclaimer was invariably lost in translation to popular accounts.) To believe, even for a moment, that such a procedure can recover the proper ordering of IQ among "men of genius," one must assume that the childhood of all subjects was watched and recorded with roughly equal attention. One must claim (as Cox does) that an absence of recorded childhood precocity indicates a humdrum life not worth writing about, not an extraordinary giftedness that no one bothered to record.

Two basic results of Cox's study immediately arouse our strong suspicion that her IQ scores reflect the historical accidents of surviving records, rather than the true accomplishments of her geniuses. First, IQ is not supposed to alter in a definite direction during a person's life. Yet average A1 IQ is 135 in her study, and average A2 IQ is a substantially higher 145. When we scrutinize her dossiers (printed in full in Cox, 1926), the reason is readily apparent, and a clear artifact of her method. She has more information on her subjects as young adults than as children (A2 IQ records achievements during ages seventeen through twenty-six; A1 IQ marks the earlier years). Second, Cox published disturbingly low A1 IQ figures for some formidable characters, including Cervantes and Copernicus, both at 105. Her dossiers show the reason: little or nothing is known about their childhood, providing no data for addition to the base figure of 100. Cox established seven levels of reliability for her figures. The seventh, believe it or not, is "guess, based on no data."

As a further and obvious test, consider geniuses born into humble circumstances, where tutors and scribes did not abound to encourage and then to record daring feats of precocity. John Stuart Mill may have learned Greek in his cradle, but did Faraday or Bunyan ever get the chance? Poor children are at a double disadvantage; not only did no one bother to record their early years, but they are also demoted as a direct result of their poverty. For Cox, using the favorite ploy of eugenicists, inferred innate parental intelligence from their occupations and social standing! She ranked parents on a scale of professions from 1 to 5, awarding their children an IQ of 100 for parental rank 3, and a bonus (or deficit) of 10 IQ points for each step above or below. A child who did nothing worth noting for the first seventeen years of his life could still score an IQ of 120 by virtue of his parent's wealth or professional standing.

Consider the case of poor Massena, Napoleon's great general, who bottomed out at 100 A1 IQ and about whom, as a child, we know nothing except that he served as a cabin boy for two long voyages on his uncle's ship. Cox writes (p. 88):

> Nephews of battleship commanders probably rate somewhat above 100 IQ; but cabin boys who remain cabin boys for two long voyages and of whom there is nothing more to report until the age of 17 than their service as cabin boys, may average below 100 IQ.

Other admirable subjects with impoverished parents and meager records should have suffered the ignominy of scores below 100. But Cox managed to fudge and temporize, pushing them all above the triple-digit divide, if only slightly. Consider the unfortunate Saint-Cyr, saved only by remote kin, and granted an A1 IQ of 105: "The father was a tanner after having been a butcher, which would give his son an occupational IQ status of 90 to 100; but two distant relatives achieved signal martial honors, thus indicating a higher strain in the family" (pp. 90–91). John Bunyan faced more familial obstacles than his famous Pilgrim, but Cox managed to extract a score of 105 for him:

> Bunyan's father was a brazier or tinker, but a tinker of recognized position in the village; and the mother was not of the squalid poor, but of people who were "decent and worthy in their ways." This would be sufficient evidence for a rating between 90 and 100. But the record goes further, and we read that notwithstanding their "meanness and inconsiderableness," Bunyan's parents put their boy to school to learn "both to read and write," which probably indicates that he showed something more than the promise of a future tinker (p. 90).

Michael Faraday squeaked by at 105, overcoming the demerit of parental standing with snippets about his reliability as an errand boy and his questioning nature. His elevated A2 IQ of 150 only records increasing information about his more notable young manhood. In one case, however, Cox couldn't bear to record the unpleasant result that her methods dictated. Shakespeare, of humble origin and unknown childhood, would have scored below 100. So Cox simply left him out, even though she included several others with equally inadequate childhood records.

Among other curiosities of scoring that reflect Cox and Terman's social prejudices, several precocious youngsters (Clive, Liebig, and Swift, in particular) were downgraded for their rebelliousness in school, particularly for their unwillingness to study classics. An animus against the performing arts is evident in the rating of composers, who (as a group) rank just above soldiers at the bottom of the final list. Consider the following understatement about Mozart (p. 129): "A child who learns to play the piano at 3, who receives and benefits by musical instruction at that age, and who studies and executes the most difficult counterpoint at age 14, is probably above the average level of his social group."

In the end, I suspect that Cox recognized the shaky basis of her work, but persisted bravely nonetheless. Correlations between rank in eminence (length of Cattell's entry) and awarded IQ were disappointing to say the least—a mere 0.25 for eminence vs. A2 IQ, with no figure recorded at all for eminence vs. A1 IQ (it is a lower 0.20 by my calculation). Instead, Cox makes much of the fact that her ten most eminent subjects average 4—yes only 4—A1 IQ points above her ten least eminent.

Cox calculated her strongest correlation (0.77) between A2 IQ and "index of reliability," a measure of available information about her subjects. I can imagine no better demonstration that Cox's IQ's are artifacts of differential amounts of data, not measures of innate ability or even, for that matter, of simple talent. Cox recognized this and, in a final effort, tried to "correct" her scores for missing information by adjusting poorly documented subjects upward toward the group means of 135 for A1 IQ and 145 for A2 IQ. These adjustments boosted average IQ's substantially, but led to other embarrassments. For uncorrected scores, the most eminent fifty averaged 142 for A1 IQ, while the least eminent fifty scored comfortably lower at 133. With corrections, the first fifty scored 160, the last fifty, 165. Ultimately, only Goethe and Voltaire scored near the top both in IQ and eminence. One might paraphrase Voltaire's famous quip about God and conclude that even though adequate information on the IQ of history's eminent men does not exist, it was probably inevitable that the American hereditarians would try to invent it.

Terman on group differences

Terman's empirical work measured what statisticians call the "within-group variance" of IQ—that is, the differences in scores within single populations (all children in a school, for example). At best, he was able to show that children testing well or poorly at a young age generally maintain their ordering with respect to other children as the population grows up. Terman ascribed most of these differences to variation in biological endowment, without much evidence beyond an assertion that all right-minded people recognize the domination of nurture by nature. This brand of hereditarianism might offend our present sensibilities with its elitism and its accompanying proposals for institutional care and forced abstinence from breeding, but it does not, by itself, entail the more contentious claim for innate differences between groups.

Terman made this invalid extrapolation, as virtually all hereditarians did and still do. He then compounded his error by confusing the genesis of true pathologies with causes for variation in normal behavior. We know, for example, that the mental retardation associated with Down's syndrome has its origin in a specific genetic defect (an extra chromosome). But we cannot therefore attribute the low IQ of many apparently normal children to an innate biology. We might as well claim that all overweight people can't help it because some very obese individuals can trace their condition to hormonal imbalances. Terman's data on the stability of ordering in IQ within groups of growing children relied largely upon the persistently low IQ of biologically afflicted individuals, despite Terman's attempt to bring all scores under the umbrella of a normal curve (1916, pp. 65–67), and thus to suggest that all variation has a common root in the possession of more or less of a single substance. In short, it is invalid to extrapolate from variation within a group to differences between groups. It is doubly invalid to use the innate biology of pathological individuals as a basis for ascribing normal variation within a group to inborn causes.

At least the IQ hereditarians did not follow their craniological forebears in harsh judgments about women. Girls did not score below boys in IQ, and Terman proclaimed their limited access to professions both unjust and wasteful of intellectual talent (1916, p. 72; 1919, p. 288). He noted, assuming that IQ should earn its monetary reward, that women scoring between 100 and 120 generally earned, as teachers or "high-grade stenographers," what men with an IQ of 85 received as motormen, firemen, or policemen (1919, p. 278).

But Terman took the hereditarian line on race and class and proclaimed its validation as a primary aim of his work. In ending his chapter on the uses of IQ (1916, pp. 19–20), Terman posed three questions:

> Is the place of the so-called lower classes in the social and industrial scale the result of their inferior native endowment, or is their apparent inferiority merely a result of their inferior home and school training? Is genius more common among children of the educated classes than among the children of the ignorant and poor? Are the inferior races really inferior, or are they merely unfortunate in their lack of opportunity to learn?

Despite a poor correlation of 0.4 between social status and IQ, Terman (1917) advanced five major reasons for claiming that "environment is much less important than is original endowment in determining the nature of the traits in question" (p. 91). The first three, based on additional correlations, add no evidence for innate causes. Terman calculated: 1) a correlation of 0.55 between social status and teachers' assessments of intelligence; 2) 0.47 between social status and school work; and 3) a lower, but unstated,* correlation between "age-grade progress" and social status. Since all five properties—IQ, social status, teacher's assessment, school work, and age-grade progress—may be redundant measures of the same complex and unknown causes, the correlation between any additional pair adds little to the basic result of 0.4 between IQ and social status. If the 0.4 correlation offers no evidence for innate causes, then the additional correlations do not either.

The fourth argument, recognized as weak by Terman himself

*It is annoyingly characteristic of Terman's work that he cites correlations when they are high and favorable, but does not give the actual figures when they are low but still favorable to his hypothesis. This ploy abounds in Cox's study of posthumous genius and in Terman's analysis of IQ among professions, both discussed previously.

(1916, p. 98), confuses probable pathology with normal variation, and is therefore irrelevant, as discussed above: feeble-minded children are occasionally born to rich or to intellectually successful parents.

The fifth argument reveals the strength of Terman's hereditarian convictions and his remarkable insensitivity to the influence of environment. Terman measured the IQ of twenty children in a California orphanage. Only three were "fully normal," while seventeen ranged from 75 to 95. The low scores cannot be attributed to life without parents, Terman argues, because (p. 99):

> The orphanage in question is a reasonably good one and affords an environment which is about as stimulating to normal mental development as average home life among the middle classes. The children live in the orphanage and attend an excellent public school in a California village.

Low scores must reflect the biology of children committed to such institutions:

> Some of the tests which have been made in such institutions indicate that mental subnormality of both high and moderate grades is extremely frequent among children who are placed in these homes. Most, though admittedly not all of these, are children of inferior social classes (p. 99).

Terman offers no direct evidence about the lives of his twenty children beyond the fact of their institutional placement. He is not even certain that they all came from "inferior social classes." Surely, the most parsimonious assumption would relate low IQ scores to the one incontestable and common fact about the children—their life in the orphanage itself.

Terman moved easily from individuals, to social classes, to races. Distressed by the frequency of IQ scores between 70 and 80, he lamented (1916, pp. 91–92):

> Among laboring men and servant girls there are thousands like them. . . . The tests have told the truth. These boys are ineducable beyond the merest rudiments of training. No amount of school instruction will ever make them intelligent voters or capable citizens. . . . They represent the level of intelligence which is very, very common among Spanish-Indian and Mexican families of the Southwest and also among negroes. Their dullness seems to be racial, or at least inherent in the family stocks from which they came. The fact that one meets this type with such extraordinary frequency among Indians, Mexicans, and negroes suggests quite forcibly

> that the whole question of racial differences in mental traits will have to be taken up anew and by experimental methods. The writer predicts that when this is done there will be discovered enormously significant racial differences in general intelligence, differences which cannot be wiped out by any scheme of mental culture. Children of this group should be segregated in special classes and be given instruction which is concrete and practical. They cannot master abstractions, but they can often be made efficient workers, able to look out for themselves. There is no possibility at present of convincing society that they should not be allowed to reproduce, although from a eugenic point of view they constitute a grave problem because of their unusually prolific breeding.

Terman sensed that his arguments for innateness were weak. Yet what did it matter? Do we need to prove what common sense proclaims so clearly?

> After all, does not common observation teach us that, in the main, native qualities of intellect and character, rather than chance, determine the social class to which a family belongs? From what is already known about heredity, should we not naturally expect to find the children of well-to-do, cultured, and successful parents better endowed than the children who have been reared in slums and poverty? An affirmative answer to the above question is suggested by nearly all the available scientific evidence (1917, p. 99).

Whose common sense?

Terman recants

Terman's book on the Stanford-Binet revision of 1937 was so different from the original volume of 1916 that common authorship seems at first improbable. But then times had changed and intellectual fashions of jingoism and eugenics had been swamped in the morass of a Great Depression. In 1916 Terman had fixed adult mental age at sixteen because he couldn't get a random sample of older schoolboys for testing. In 1937 he could extend his scale to age eighteen; for "the task was facilitated by the extremely unfavorable employment situation at the time the tests were made, which operated to reduce considerably the school elimination normally occurring after fourteen" (1937, p. 30).

Terman did not explicity abjure his previous conclusions, but a veil of silence descended upon them. Not a word beyond a few statements of caution do we hear about heredity. All potential reasons for differences between groups are framed in environmental terms. Terman presents his old curves for average differences in IQ between social classes, but he warns us that mean differences are too small to provide any predictive information for individuals. We also do not know how to partition the average differences between genetic and environmental influences:

> It is hardly necessary to stress the fact that these figures refer to mean values only, and that in view of the variability of the IQ within each group the respective distributions greatly overlap one another. Nor should it be necessary to point out that such data do not, in themselves, offer any conclusive evidence of the relative contributions of genetic and environmental factors in determining the mean differences observed.

A few pages later, Terman discusses the differences between rural and urban children, noting the lower country scores and the curious finding that rural IQ drops with age after entrance to school, while IQ for urban children of semiskilled and unskilled workers rises. He expresses no firm opinion, but note that the only hypotheses he wishes to test are now environmental:

> It would require extensive research, carefully planned for the purpose, to determine whether the lowered IQ of rural children can be ascribed to the relatively poorer educational facilities in rural communities, and whether the gain for children from the lower economic strata can be attributed to an assumed enrichment of intellectual environment that school attendance bestows.

Autres temps, autres moeurs.

R. M. Yerkes and the Army Mental Tests: IQ comes of age

Psychology's great leap forward

Robert M. Yerkes, about to turn forty, was a frustrated man in 1915. He had been on the faculty of Harvard University since 1902. He was a superb organizer of men, and an eloquent promotor of his profession. Yet psychology still wallowed in its reputation as a "soft" science, if a science at all. Some colleges did not acknowledge its existence; others ranked it among the humanities and placed psychologists in departments of philosophy. Yerkes wished, above all, to establish his profession by proving that it could be as rigorous a science as physics. Yerkes and most of his contemporaries equated rigor and science with numbers and quantification. The most promising source of copious and objective numbers, Yerkes believed, lay in the embryonic field of mental testing. Psychology would come of age, and gain acceptance as a true science worthy of financial and institutional support, if it could bring the question of human potential under the umbrella of science:

> Most of us are wholly convinced that the future of mankind depends in no small measure upon the development of the various biological and social sciences. . . . We must . . . strive increasingly for the improvement of our methods of mental measurement, for there is no longer ground for doubt concerning the practical as well as the theoretical importance of studies of human behavior. We must learn to measure skillfully every form and aspect of behavior which has psychological and sociological significance (Yerkes, 1917a, p. 111).

But mental testing suffered from inadequate support and its own internal contradictions. It was, first of all, practiced extensively by poorly trained amateurs whose manifestly absurd results were giving the enterprise a bad name. In 1915, at the annual meeting of the American Psychological Association in Chicago, a critic reported that the mayor of Chicago himself had tested as a moron on one version of the Binet scales. Yerkes joined with critics in discussions at the meeting and proclaimed: "We are building up a science, but we have not yet devised a mechanism which anyone can operate" (quoted in Chase, 1977, p. 242).

Second, available scales gave markedly different results even when properly applied. As discussed on p. 166, half the individuals who tested in the low, but normal range on the Stanford-Binet, were morons on Goddard's version of the Binet scale. Finally, support had been too inadequate, and coordination too sporadic, to build up a pool of data sufficiently copious and uniform to compel belief (Yerkes, 1917b).

Wars always generate their retinue of camp followers with ulterior motives. Many are simply scoundrels and profiteers, but a few are spurred by higher ideals. As mobilization for World War I approached, Yerkes got one of those "big ideas" that propel the history of science: could psychologists possibly persuade the army to test all its recruits? If so, the philosopher's stone of psychology might be constructed: the copious, useful, and uniform body of numbers that would fuel a transition from dubious art to respected science. Yerkes proselytized within his profession and within government circles, and he won his point. As Colonel Yerkes, he presided over the administration of mental tests to 1.75 million recruits during World War I. Afterward, he proclaimed that mental testing "helped to win the war." "At the same time," he added, "it has incidentally established itself among the other sciences and demonstrated its right to serious consideration in human engineering" (quoted in Kevles, 1968, p. 581).

Yerkes brought together all the major hereditarians of American psychometrics to write the army mental tests. From May to July 1917 he worked with Terman, Goddard, and other colleagues at Goddard's Training School in Vineland, New Jersey.

Their scheme included three types of tests. Literate recruits would be given a written examination, called the Army Alpha. Illiterates and men who had failed Alpha would be given a pictorial test, called the Army Beta. Failures in Beta would be recalled for an individual examination, usually some version of the Binet scales. Army psychologists would then grade each man from A to E (with plusses and minuses) and offer suggestions for proper military placement. Yerkes suggested that recruits with a score of C− should be marked as "low average intelligence—ordinary private." Men of grade D are "rarely suited for tasks requiring special skill, forethought, resourcefulness or sustained alertness." D and E men could not be expected "to read and understand written directions."

I do not think that the army ever made much use of the tests. One can well imagine how professional officers felt about smart-assed young psychologists who arrived without invitation, often assumed an officer's rank without undergoing basic training, commandeered a building to give the tests (if they could), saw each recruit for an hour in a large group, and then proceeded to usurp an officer's traditional role in judging the worthiness of men for various military tasks. Yerkes's corps encountered hostility in some camps; in others, they suffered a penalty in many ways more painful: they were treated politely, given appropriate facilities, and then ignored.* Some army officials became suspicious of Yerkes's

*Yerkes continued to complain throughout his career that military psychology had not achieved its due respect, despite its accomplishments in World War I. During World War II the aging Yerkes was still grousing and arguing that the Nazis were

intent and launched three independent investigations of the testing program. One concluded that it should be controlled so that "no theorist may . . . ride it as a hobby for the purpose of obtaining data for research work and the future benefit of the human race" (quoted in Kevles, 1968, p. 577).

Still, the tests did have a strong impact in some areas, particularly in screening men for officer training. At the start of the war, the army and national guard maintained nine thousand officers. By the end, two hundred thousand officers presided, and two-thirds of them had started their careers in training camps where the tests were applied. In some camps, no man scoring below C could be considered for officer training.

But the major impact of Yerkes's tests did not fall upon the army. Yerkes may not have brought the army its victory, but he certainly won his battle. He now had uniform data on 1.75 million men, and he had devised, in the Alpha and Beta exams, the first mass-produced written tests of intelligence. Inquiries flooded in from schools and businesses. In his massive monograph (Yerkes, 1921) on *Psychological Examining in the United States Army,* Yerkes buried a statement of great social significance in an aside on page 96. He spoke of "the steady stream of requests from commercial concerns, educational institutions, and individuals for the use of army methods of psychological examining or for the adaptation of such methods to special needs." Binet's purpose could now be circumvented because a technology had been developed for testing all pupils. Tests could now rank and stream everybody; the era of mass testing had begun.

Results of the army tests

The primary impact of the tests arose not from the army's lackadaisical use of scores for individuals, but from general propaganda that accompanied Yerkes's report of the summary statistics (Yerkes, 1921, pp. 553–875). E. G. Boring, later a famous psychol-

---

upstaging America in their proper use and encouragement of mental testing for military personnel. "Germany has a long lead in the development of military psychology. . . . The Nazis have achieved something that is entirely without parallel in military history. . . . What has happened in Germany is the logical sequel to the psychological and personnel services in our own Army during 1917–1918" (Yerkes, 1941, p. 209).

ogist himself but then Yerkes's lieutenant (and the army's captain), selected one hundred sixty thousand cases from the files and produced data that reverberated through the 1920s with a hard hereditarian ring. The task was a formidable one. The sample, which Boring culled himself with the aid of only one assistant, was very large; moreover, the scales of three different tests (Alpha, Beta, and individual) had to be converted to a common standard so that racial and national averages could be constructed from samples of men who had taken the tests in different proportions (few blacks took Alpha, for example).

From Boring's ocean of numbers, three "facts" rose to the top and continued to influence social policy in America long after their source in the tests had been forgotten.

1. The average mental age of white American adults stood just above the edge of moronity at a shocking and meager thirteen. Terman had previously set the standard at sixteen. The new figure became a rallying point for eugenicists who predicted doom and lamented our declining intelligence, caused by the unconstrained breeding of the poor and feeble-minded, the spread of Negro blood through miscegenation, and the swamping of an intelligent native stock by the immigrating dregs of southern and eastern Europe. Yerkes* wrote:

> It is customary to say that the mental age of the average adult is about 16 years. This figure is based, however, upon examinations of only 62 persons; 32 of them high-school pupils from 16–20 years of age, and 30 of them "business men of moderate success and of very limited educational advantages." The group is too small to give very reliable results and is furthermore probably not typical. . . . It appears that the intelligence of the principal sample of the white draft, when transmuted from Alpha and Beta exams into terms of mental age, is about 13 years (13.08) (1921, p. 785).

Yet, even as he wrote, Yerkes began to sense the logical absurdity of such a statement. An average is what it is; it cannot lie three years below what it should be. So Yerkes thought again and added:

> We can hardly say, however, with assurance that these recruits are three years mental age below the average. Indeed, it might be argued on

*I doubt that Yerkes wrote all parts of the massive 1921 monograph himself. But he is listed as the only author of this official report, and I shall continue to attribute its statements to him, both as shorthand and for want of other information.

extrinsic grounds that the draft itself is more representative of the average intelligence of the country than is a group of high-school students and business men (1921, p. 785).

If 13.08 is the white average, and everyone from mental age 8 through 12 is a moron, then we are a nation of nearly half-morons. Yerkes concluded (1921, p. 791): "It would be totally impossible to exclude all morons as that term is at present defined, for there are under 13 years 37 percent of whites and 89 percent of negroes."

2. European immigrants can be graded by their country of origin. The average man of many nations is a moron. The darker peoples of southern Europe and the Slavs of eastern Europe are less intelligent than the fair peoples of western and northern Europe. Nordic supremacy is not a jingoistic prejudice. The average Russian has a mental age of 11.34; the Italian, 11.01; the Pole, 10.74. The Polish joke attained the same legitimacy as the moron joke—indeed, they described the same animal.

3. The Negro lies at the bottom of the scale with an average mental age of 10.41. Some camps tried to carry the analysis a bit further, and in obvious racist directions. At Camp Lee, blacks were divided into three groups based upon intensity of color; the lighter groups scored higher (p. 531). Yerkes reported that the opinions of officers matched his numbers (p. 742):

> All officers without exception agree that the negro lacks initiative, displays little or no leadership, and cannot accept responsibility. Some point out that these defects are greater in the southern negro. All officers seem further to agree that the negro is a cheerful, willing soldier, naturally subservient. These qualities make for immediate obedience, although not necessarily for good discipline, since petty thieving and venereal disease are commoner than with white troops.

Along the way, Yerkes and company tested several other social prejudices. Some fared poorly, particularly the popular eugenical notion that most offenders are feeble-minded. Among conscientious objectors for political reasons, 59 percent received a grade of A. Even outright disloyals scored above the average (p. 803). But other results buoyed their prejudices. As camp followers themselves, Yerkes's corps decided to test a more traditional category of colleagues: the local prostitutes. They found that 53 percent (44 percent of whites and 68 percent of blacks) ranked at age ten or below on the Goddard version of the Binet scales. (They acknowledge that the Goddard scales ranked people well below their scores on other versions of the Binet tests.) Yerkes concluded (p. 808):

> The results of Army examining of prostitutes corroborate the conclusion, attained by civilian examinations of prostitutes in various parts of the country, that from 30 to 60 percent of prostitutes are deficient and are for the most part high-grade morons; and that 15 to 25 percent of all prostitutes are so low-grade mentally that it is wise (as well as possible under the existing laws in most states) permanently to segregate them in institutions for the feeble-minded.

One must be thankful for small bits of humor to lighten the reading of an eight-hundred-page statistical monograph. The thought of army personnel rounding up the local prostitutes and sitting them down to take the Binet tests amused me no end, and must have bemused the ladies even more.

As pure numbers, these data carried no inherent social message. They might have been used to promote equality of opportunity and to underscore the disadvantages imposed upon so many Americans. Yerkes might have argued that an average mental age of thirteen reflected the fact that relatively few recruits had the opportunity to finish or even to attend high school. He might have attributed the low average of some national groups to the fact that most recruits from these countries were recent immigrants who did not speak English and were unfamiliar with American culture. He might have recognized the link between low Negro scores and the history of slavery and racism.

But scarcely a word do we read through eight hundred pages of any role for environmental influence. The tests had been written by a committee that included all the leading American hereditarians discussed in this chapter. They had been constructed to measure innate intelligence, and they did so by definition. The circularity of argument could not be broken. All the major findings received hereditarian interpretations, often by near miracles of special pleading to argue past a patent environmental influence. A circular issued from the School of Military Psychology at Camp Greenleaf proclaimed (do pardon its questionable grammar): "These tests do not measure occupational fitness nor educational attainment; they measure intellectual ability. This latter has been shown to be important in estimating military value" (p. 424). And the boss himself argued (Yerkes, quoted in Chase, 1977, p. 249):

> Examinations Alpha and Beta are so constructed and administered as to minimize the handicap of men who because of foreign birth or lack of education are little skilled in the use of English. These group examinations were originally intended, and are now definitely known, to measure native intellectual ability. They are to some extent influenced by educational acquirement, but in the main the soldier's inborn intelligence and not the accidents of environment determines his mental rating or grade in the army.

A critique of the Army Mental Tests

THE CONTENT OF THE TESTS

The Alpha test included eight parts, the Beta seven; each took less than an hour and could be given to large groups. Most of the Alpha parts presented items that have become familiar to generations of test-takers ever since: analogies, filling in the next number in a sequence, unscrambling sentences, and so forth. This similarity is no accident; the Army Alpha was the granddaddy, literally as well as figuratively, of all written mental tests. One of Yerkes's disciples, C. C. Brigham, later became secretary of the College Entrance Examination Board and developed the Scholastic Aptitude Test on army models. If people get a peculiar feeling of déjà-vu in perusing Yerkes's monograph, I suggest that they think back to their own College Boards, with all its attendant anxiety.

These familiar parts are not especially subject to charges of cultural bias, at least no more so than their modern descendants. In a general way, of course, they test literacy, and literacy records education more than inherited intelligence. Moreover, a schoolmaster's claim that he tests children of the same age and school experience, and therefore may be recording some internal biology, didn't apply to the army recruits—for they varied greatly in access to education and recorded different amounts of schooling in their scores. A few of the items are amusing in the light of Yerkes's assertion that the tests "measure native intellectual ability." Consider the Alpha analogy: "Washington is to Adams as first is to. . . ."

But one part of each test is simply ludicrous in the light of Yerkes's analysis. How could Yerkes and company attribute the low scores of recent immigrants to innate stupidity when their multiple-choice test consisted entirely of questions like:

> Crisco is a: patent medicine, disinfectant, toothpaste, food product
> The number of a Kaffir's legs is: 2, 4, 6, 8
> Christy Mathewson is famous as a: writer, artist, baseball player, comedian

I got the last one, but my intelligent brother, who, to my distress, grew up in New York utterly oblivious to the heroics of three great baseball teams then resident, did not.

Yerkes might have responded that recent immigrants generally took Beta rather than Alpha, but Beta contains a pictorial version of the same theme. In this complete-a-picture test, early items might be defended as sufficiently universal: adding a mouth to a face or an ear to a rabbit. But later items required a rivet in a pocket knife, a filament in a light bulb, a horn on a phonograph, a net on a tennis court, and a ball in a bowler's hand (marked wrong, Yerkes explained, if an examinee drew the ball in the alley, for you can tell from the bowler's posture that he has not yet released the ball). Franz Boas, an early critic, told the tale of a Sicilian recruit who added a crucifix where it always appeared in his native land to a house without a chimney. He was marked wrong.

The tests were strictly timed, for the next fifty were waiting by the door. Recruits were not expected to finish each part; this was explained to the Alpha men, but not to Beta people. Yerkes wondered why so many recruits scored flat zero on so many of the parts (the most telling proof of the tests' worthlessness—see pp. 213–214). How many of us, if nervous, uncomfortable, and crowded (and even if not), would have understood enough to write anything at all in the ten seconds allotted for completing the following commands, each given but once in Alpha, Part 1?

> Attention! Look at 4. When I say "go" make a figure 1 in the space which is in the circle but not in the triangle or square, and also make a figure 2 in the space which is in the triangle and circle, but not in the square. Go.
>
> Attention! Look at 6. When I say "go" put in the second circle the right answer to the question: "How many months has a year?" In the third circle do nothing, but in the fourth circle put any number that is a wrong answer to the question that you have just answered correctly. Go.

INADEQUATE CONDITIONS

Yerkes's protocol was rigorous and trying enough. His examiners had to process men rapidly and grade the exams immediately, so that failures could be recalled for a different test. When faced with the added burden of thinly veiled hostility from the brass at several camps, Yerkes's testers were rarely able to carry out more than a caricature of their own stated procedure. They continually compromised, backtracked, and altered in the face of necessity. Procedures varied so much from camp to camp that results could scarcely be collated and compared. The whole effort, through no fault of Yerkes's beyond impracticality and overambition, became something of a shambles, if not a disgrace. The details are all in Yerkes's monograph, but hardly anyone ever read it. The summary statistics became an important social weapon for racists and eugenicists; their rotten core lay exposed in the monograph, but who looks within when the surface shines with such a congenial message.

The army mandated that special buildings be supplied or even constructed for Yerkes's examinations, but a different reality prevailed (1921, p. 61). The examiners had to take what they could get, often rooms in cramped barracks with no furnishings at all, and inadequate acoustics, illumination, and lines of sight. The chief tester at one camp complained (p. 106): "Part of this inaccuracy I believe to be due to the fact that the room in which the examination is held is filled too full of men. As a result, the men who are sitting in the rear of the room are unable to hear clearly and thoroughly enough to understand the instructions."

Tensions rose between Yerkes's testers and regular officers. The chief tester of Camp Custer complained (p. 111): "The ignorance of the subject on the part of the average officer is equalled only by his indifference to it." Yerkes urged restraint and accommodation (p. 155):

> The examiner should strive especially to take the military point of view. Unwarranted claims concerning the accuracy of the results should be avoided. In general, straightforward commonsense statements will be found more convincing than technical descriptions, statistical exhibits, or academic arguments.

As friction and doubt mounted, the secretary of war polled commanding officers of all camps to ask their opinion of Yerkes's tests. He received one hundred replies, nearly all negative. They were, Yerkes admitted (p. 43), "with a few exceptions, unfavorable to psychological work, and have led to the conclusion on the part of various officers of the General Staff that this work has little, if any, value to the army and should be discontinued." Yerkes fought back and won a standoff (but not all the promotions, commissions, and hirings he had been promised); his work proceeded under a cloud of suspicion.

Minor frustrations never abated. Camp Jackson ran out of forms and had to improvise on blank paper (p. 78). But a major and persistent difficulty dogged the entire enterprise and finally, as I shall demonstrate, deprived the summary statistics of any meaning. Recruits had to be allocated to their appropriate test. Men illiterate in English, either by lack of schooling or foreign birth, should have taken examination Beta, either by direct assignment, or indirectly upon failing Alpha. Yerkes's corps tried heroically to fulfill this procedure. In at least three camps, they marked identification tags or even painted letters directly on the bodies of men who failed—a ready identification guide for further assessment (p. 73, p. 76): "A list of D men was sent within six hours after the group examination to the clerk at the mustering office. As the men appeared, this clerk marked on the body of each D man a letter P" (indicating that the psychiatrist should examine them further).

But standards for the division between Alpha and Beta varied substantially from camp to camp. A survey across camps revealed that the minimum score on an early version of Alpha varied from 20 to 100 for assignment to further testing (p. 476). Yerkes admitted (p. 354):

> This lack of a uniform process of segregation is certainly unfortunate. On account of the variable facilities for examining and the variable quality of the groups examined however, it appeared entirely impossible to establish a standard uniform for all camps.

C. C. Brigham, Yerkes's most zealous votary, even complained (1921):

> The method of selecting men for Beta varied from camp to camp, and sometimes from week to week in the same camp. There was no established criterion of literacy, and no uniform method of selecting illiterates.

The problem cut far deeper than simple inconsistency among camps. The persistent logistical difficulties imposed a systematic bias that substantially lowered the mean scores of blacks and immigrants. For two major reasons, many men took only Alpha and scored either zero or next to nothing, not because they were innately dumb, but because they were illiterate and should have taken Beta by Yerkes's own protocol. First, recruits and draftees had, on average, spent fewer years in school than Yerkes had anticipated. Lines for Beta began to lengthen and the entire operation threatened to clog at this bottleneck. At many camps, unqualified men were sent in droves to Alpha by artificial lowering of standards. Schooling to the third grade sufficed for Alpha in one camp; in another, anyone who said he could read, at whatever level, took Alpha. The chief tester at Camp Dix reported (p. 72): "To avoid excessively large Beta groups, standards for admission to examination Alpha were set low."

Second, and more important, the press of time and the hostility of regular officers often precluded a Beta retest for men who had incorrectly taken Alpha. Yerkes admitted (p. 472): "It was never successfully shown, however, that the continued recalls . . . were so essential that repeated interference with company maneuvers should be permitted." As the pace became more frantic, the problem worsened. The chief tester at Camp Dix complained (pp. 72–73): "In June it was found impossible to recall a thousand men listed for individual examination. In July Alpha failures among negroes were not recalled." The stated protocol scarcely applied to blacks who, as usual, were treated with less concern and more contempt by everyone. Failure on Beta, for example, should have led to an individual examination. Half the black recruits scored D− on Beta, but only one-fifth of these were recalled and four-fifths received no further examination (p. 708). Yet we know that scores for blacks improved substantially when the protocol was followed. At one camp (p. 736), only 14.1 percent of men who had scored D− on Alpha failed to gain a higher grade on Beta.

The effects of this systematic bias are evident in one of Boring's experiments with the summary statistics. He culled 4,893 cases of men who had taken both Alpha and Beta. Converting their scores to the common scale, he calculated an average mental age of 10.775 for Alpha, and a Beta mean of 12.158 (p. 655). He used only the Beta scores in his summaries; Yerkes procedure worked. But what of the myriads who should have taken Beta, but only received Alpha and scored abysmally as a result—primarily poorly educated blacks and immigrants with an imperfect command of English—the very groups whose low scores caused such a hereditarian stir later on?

DUBIOUS AND PERVERSE PROCEEDINGS: A PERSONAL TESTIMONY

Academicians often forget how poorly or incompletely the written record, their primary source, may represent experience. Some things have to be seen, touched, and tasted. What was it like to be an illiterate black or foreign recruit, anxious and befuddled at the novel experience of taking an examination, never told why, or what would be made of the results: expulsion, the front lines? In 1968 (quoted in Kevles), an examiner recalled his administration of Beta: "It was touching to see the intense effort . . . put into answering the questions, often by men who never before had held a pencil in their hands." Yerkes had overlooked, or consciously bypassed, something of importance. The Beta examination contained only pictures, numbers, and symbols. But it still required pencil work and, on three of its seven parts, a knowledge of numbers and how to write them.

Yerkes's monograph is so thorough that his procedure for giving the two examinations can be reconstructed down to the choreography of motion for all examiners and orderlies. He provides facsimiles in full size for the examinations themselves, and for all explanatory material used by examiners. The standardized words and gestures of examiners are reproduced in full. Since I wanted to know in as complete a way as possible what it felt like to give and take the test, I administered examination Beta (for illiterates) to a group of fifty-three Harvard undergraduates in my course on biology as a social weapon. I tried to follow Yerkes's protocol scrupulously in all its details. I feel that I reconstructed the original situation accurately, with one important exception: my students knew what they were doing, didn't have to provide their names on the form, and had nothing at stake. (One friend later suggested that I should have required names—and posted results—as just a small contribution to simulating the anxiety of the original.)

I knew before I started that internal contradictions and a priori prejudice thoroughly invalidated the hereditarian conclusions that Yerkes had drawn from the results. Boring himself called these conclusions "preposterous" late in his career (in a 1962 interview, quoted in Kevles, 1968). But I had not understood how the Draconian conditions of testing made such a thorough mockery of the claim that recruits could have been in a frame of mind to record anything about their innate abilities. In short, most of the men must have ended up either utterly confused or scared shitless.

The recruits were ushered into a room and seated before an examiner and demonstrator standing atop a platform, and several orderlies at floor level. Examiners were instructed to administer the test "in a genial manner" since "the subjects who take this examination sometimes sulk and refuse to work" (p. 163). Recruits were told nothing about the examination or its purposes. The examiner simply said: "Here are some papers. You must not open them or turn them over until you are told to." The men then filled in their names, age, and education (with help for those too illiterate to do so). After these perfunctory preliminaries, the examiner plunged right in:

> Attention. Watch this man (pointing to demonstrator). He (pointing to demonstrator again) is going to do here (tapping blackboard with pointer) what you (pointing to different members of the group) are to do on your papers (here examiner points to several papers that lie before men in the group, picks up one, holds it next to the blackboard, returns the paper, points to demonstrator and the blackboard in succession, then to the men and their papers). Ask no questions. Wait till I say "Go ahead!" (p. 163).

By comparison, Alpha men were virtually inundated with information (p. 157), for the Alpha examiner said:

> Attention! The purpose of this examination is to see how well you can remember, think, and carry out what you are told to do. We are not looking for crazy people. The aim is to help find out what you are best fitted to do in the Army. The grade you make in this examination will be put on your qualification card and will also go to your company commander. Some of the things you are told to do will be very easy. Some you may find

hard. You are not expected to make a perfect grade, but do the very best you can. . . . Listen closely. Ask no questions.

The extreme limits imposed upon the Beta examiner's vocabulary did not only reflect Yerkes's poor opinion of what Beta recruits might understand by virtue of their stupidity. Many Beta examinees were recent immigrants who did not speak English, and instruction had to be as pictorial and gestural as possible. Yerkes advised (p. 163): "One camp has had great success with a 'window seller' as demonstrator. Actors should also be considered for the work." One particularly important bit of information was not transmitted: examinees were not told that it was virtually impossible to finish at least three of the tests, and that they were not expected to do so.

Atop the platform, the demonstrator stood in front of a blackboard roll covered by a curtain; the examiner stood at his side. Before each of the seven tests, the curtain was raised to expose a sample problem (all reproduced in Figure 5.4), and examiner and demonstrator engaged in a bit of pantomime to illustrate proper procedure. The examiner then issued an order to work, and the demonstrator closed the curtain and advanced the roll to the next sample. The first test, maze running, received the following demonstration:

Demonstrator traces path through first maze with crayon, slowly and hesitatingly. Examiner then traces second maze and motions to demonstrator to go ahead. Demonstrator makes one mistake by going into the blind alley at upper left-hand corner of maze. Examiner apparently does not notice what demonstrator is doing until he crosses line at end of alley; then examiner shakes his head vigorously, says "No-no," takes demonstrator's hand and traces back to the place where he may start right again. Demonstrator traces rest of maze so as to indicate an attempt at haste, hesitating only at ambiguous points. Examiner says "Good." Then holding up blank, "Look here," and draws an imaginary line across the page from left to right for every maze on the page. Then, "All right. Go ahead. Do it (pointing to men and then to books). Hurry up."

This paragraph may be naively amusing (some of my students thought so). The next statement, by comparison, is a bit diabolical.

The idea of working fast must be impressed on the men during the maze test. Examiner and orderlies walk around the room, motioning to men who are not working, and saying, "Do it, do it, hurry up, quick." At the end of 2 minutes examiner says, "Stop! Turn over the page to test 2."

The examiner demonstrated test 2, cube counting, with three-dimensional models (my son had some left over from his baby days). Note that recruits who could not write numbers would receive scores of zero even if they counted all the cubes correctly. Test 3, the X-O series, will be recognized by nearly everyone today as the pictorial version of "what is the next number in the sequence." Test 4, digit symbols, required the translation of nine digits into corresponding symbols. It looks easy enough, but the test itself included ninety items and could hardly be finished by anybody in the two minutes allotted. A man who couldn't write numbers was faced with two sets of unfamiliar symbols and suffered a severe additional disadvantage. Test 5, number checking, asked men to compare numerical sequences, up to eleven digits in length, in two parallel columns. If items on the same line were identical in the two columns, recruits were instructed (by gestures) to write an X next to the item. Fifty sequences occupied three minutes, and few recruits could finish. Again, an inability to write or recognize numbers would make the task virtually impossible.

Test 6, pictorial completion, is Beta's visual analogue of Alpha's multiple-choice examination for testing innate intelligence by asking recruits about commercial products, famous sporting or film stars, or the primary industries of various cities and states. Its instructions are worth repeating:

"This is test 6 here. Look. A lot of pictures." After everyone has found the place, "Now watch." Examiner points to hand and says to demonstrator, "Fix it." Demonstrator does nothing, but looks puzzled. Examiner points to the picture of the hand, and then to the place where the finger is missing and says to demonstrator, "Fix it; fix it." Demonstrator then draws in finger. Examiner says, "That's right." Examiner then points to fish and place for eye and says, "Fix it." After demonstrator has drawn missing eye, examiner points to each of the four remaining drawings and says, "Fix them all." Demonstrator works samples out slowly and with apparent effort. When the samples are finished examiner says, "All right. Go ahead. Hurry up!" During the course of this test the orderlies walk around the room and locate individuals who are doing nothing, point to their pages and say, "Fix it. Fix them," trying to set everyone working. At the end of 3 minutes examiner says, "Stop! But don't turn over the page."

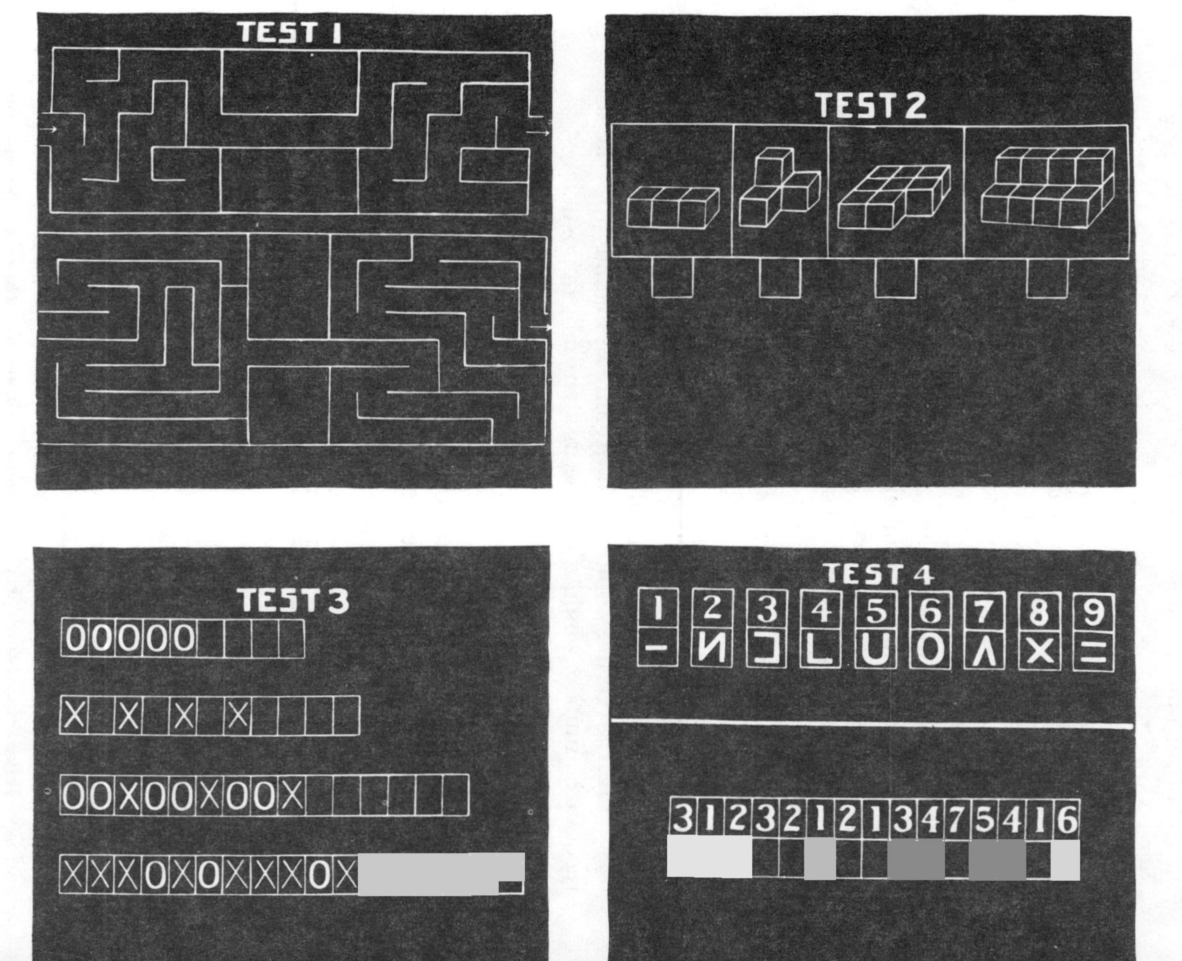
TEST 1
TEST 2
TEST 3
OOOOO
X X X X
OOXOOXOOX
XXXOXOXXXOX
TEST 4
1 2 3 4 5 6 7 8 9
31232121347541 6

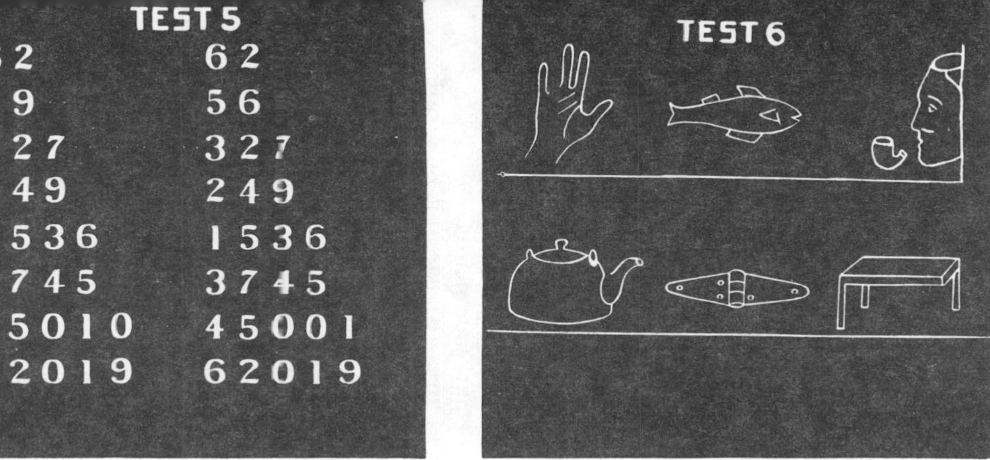

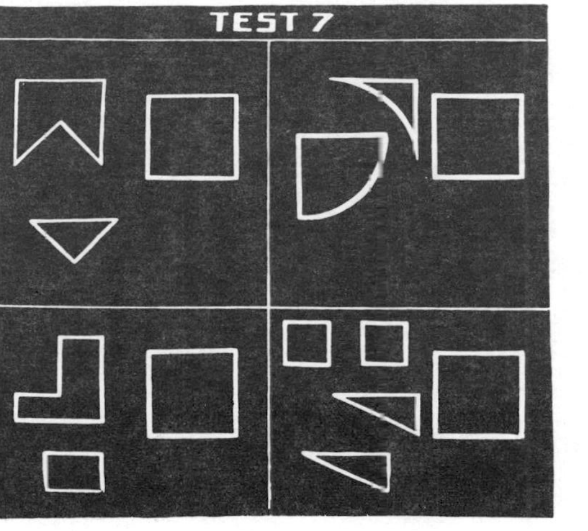

5•4 The blackboard demonstrations for all seven parts of the Beta test. From Yerkes, 1921.

The examination itself is also worth reprinting (Fig. 5.5). Best of luck with pig tails, crab legs, bowling balls, tennis nets, and the Jack's missing diamond, not to mention the phonograph horn (a real stumper for my students). Yerkes provided the following instructions for grading:

Rules for Individual Items

Item 4.—Any spoon at any angle in right hand receives credit. Left hand, or unattached spoon, no credit.
Item 5.—Chimney must be in right place. No credit for smoke.
Item 6.— Another ear on same side as first receives no credit.
Item 8.—Plain square, cross, etc., in proper location for stamp, receives credit.
Item 10.—Missing part is the rivet. Line of "ear" may be omitted.
Item 13.—Missing part is leg.
Item 15.—Ball should be drawn in hand of man. If represented in hand of woman, or in motion, no credit.
Item 16.—Single line indicating net receives credit.
Item 18.—Any representation intended for horn, pointing in any direction, receives credit.
Item 19.—Hand and powder puff must be put on proper side.
Item 20.—Diamond is the missing part. Failure to complete hilt on sword is not an error.

The seventh and last test, geometrical construction, required that a square be broken into component pieces. Its ten parts were allotted two and a half minutes.

I believe that the conditions of testing, and the basic character of the examination, make it ludicrous to believe that Beta measured any internal state deserving the label intelligence. Despite the plea for geniality, the examination was conducted in an almost frantic rush. Most parts could not be finished in the time allotted, but recruits were not forewarned. My students compiled the following record of completions on the seven parts (see p. 212). For two of the tests, digit symbols and number checking (4 and 5), most students simply couldn't write fast enough to complete the ninety and fifty items, even though the protocol was clear to all. The third test with a majority of incompletes, cube counting (number 2), was too difficult for the number of items included and the time allotted.

In summary, many recruits could not see or hear the examiner;

5 • 5 Part six of examination Beta for testing innate intelligence.

TEST	FINISHED	NOT FINISHED
1	44	9
2	21	32
3	45	8
4	12	41
5	18	35
6	49	4
7	40	13

some had never taken a test before or even held a pencil. Many did not understand the instructions and were completely befuddled. Those who did comprehend could complete only a small part of most tests in the allotted time. Meanwhile, if anxiety and confusion had not already reached levels sufficiently high to invalidate the results, the orderlies continually marched about, pointing to individual recruits and ordering them to hurry in voices loud enough, as specifically mandated, to convey the message generally. Add to this the blatant cultural biases of test 6, and the more subtle biases directed against those who could not write numbers or who had little experience in writing anything at all, and what do you have but a shambles.

The proof of inadequacy lies in the summary statistics, though Yerkes and Boring chose to interpret them differently. The monograph presents frequency distributions for scores on each part separately. Since Yerkes believed that innate intelligence was normally distributed (the "standard" pattern with a single mode at some middle score and symmetrically decreasing frequencies away from the mode in both directions), he expected that scores for each test would be normally distributed as well. But only two of the tests, maze running and picture completion (1 and 6), yielded a distribution even close to normal. (These are also the tests that my own students found easiest and completed in highest proportion.) All the other tests yielded a bimodal distribution, with one peak at a middle value and another squarely at the minimum value of zero (Fig. 5.6).

The common-sense interpretation of this bimodality holds that recruits had two different responses to the tests. Some understood what they were supposed to do, and performed in varied ways.

Others, for whatever reasons, could not fathom the instructions and scored zero. With high levels of imposed anxiety, poor conditions for seeing and hearing, and general inexperience with testing for most recruits, it would be fatuous to interpret the zero scores as evidence of innate stupidity below the intelligence of men who made some points—though Yerkes wormed out of the difficulty this way (see pp. 213–214). (My own students compiled lowest rates of completion for the tests that yield the largest secondary modes at zero in Yerkes's sample—tests 4 and 5. As the only exception to this pattern, most of my students completed test 3, which produced a strong zero mode in the army sample. But 3 is the visual analog of "what is the next number in this series," a test that all my

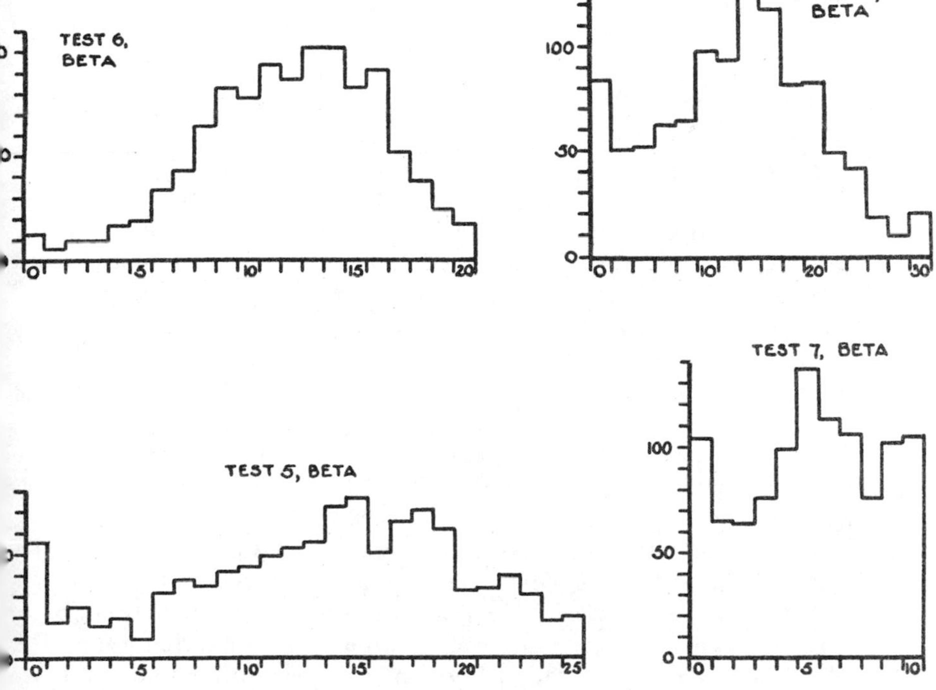

5 • 6 Frequency distributions for four of the Beta tests. Note the prominent mode at zero for tests 4, 5, and 7.

students have taken more times than they care to remember.)

Statisticians are trained to be suspicious of distributions with multiple modes. Such distributions usually indicate inhomogeneity in the system, or, in plainer language, different causes for the different modes. All familiar proverbs about the inadvisibility of mixing apples and oranges apply. The multiple modes should have guided Yerkes to a suspicion that his tests were not measuring a single entity called intelligence. Instead, his statisticians found a way to redistribute zero scores in a manner favorable to hereditarian assumptions (see next section).

Oh yes, was anyone wondering how my students fared? They did very well of course. Anything else would have been shocking, since all the tests are greatly simplified precursors of examinations they have been taking all their lives. Of fifty-three students, thirty-one scored A and sixteen B. Still, more than 10 percent (six of fifty-three) scored at the intellectual borderline of C; by the standards of some camps, they would have been fit only for the duties of a buck private.

FINAGLING THE SUMMARY STATISTICS: THE PROBLEM OF ZERO VALUES

If the Beta test faltered on the artifact of a secondary mode for zero scores, the Alpha test became an unmitigated disaster for the same reason, vastly intensified. The zero modes were pronounced in Beta, but they never reached the height of the primary mode at a middle value. But six of eight Alpha tests yielded their highest mode at zero. (Only one had a normal distribution with a middle mode, while the other yielded a zero mode lower than the middle mode.) The zero mode often soared above all other values. In one test, nearly 40 percent of all scores were zero (Fig. 5.7a). In another, zero was the only common value, with a flat distribution of other scores (at about one-fifth the level of zero values) until an even decline began at high scores (Fig. 5.7b).

Again, the common-sense interpretation of numerous zeros suggests that many men didn't understand the instructions and that the tests were invalid on that account. Buried throughout Yerkes's monograph are numerous statements proving that testers worried greatly about the high frequency of zeros and, in the midst

5 • 7a, 5 • 7b Zero was by far the most common value in several of the Alpha tests.

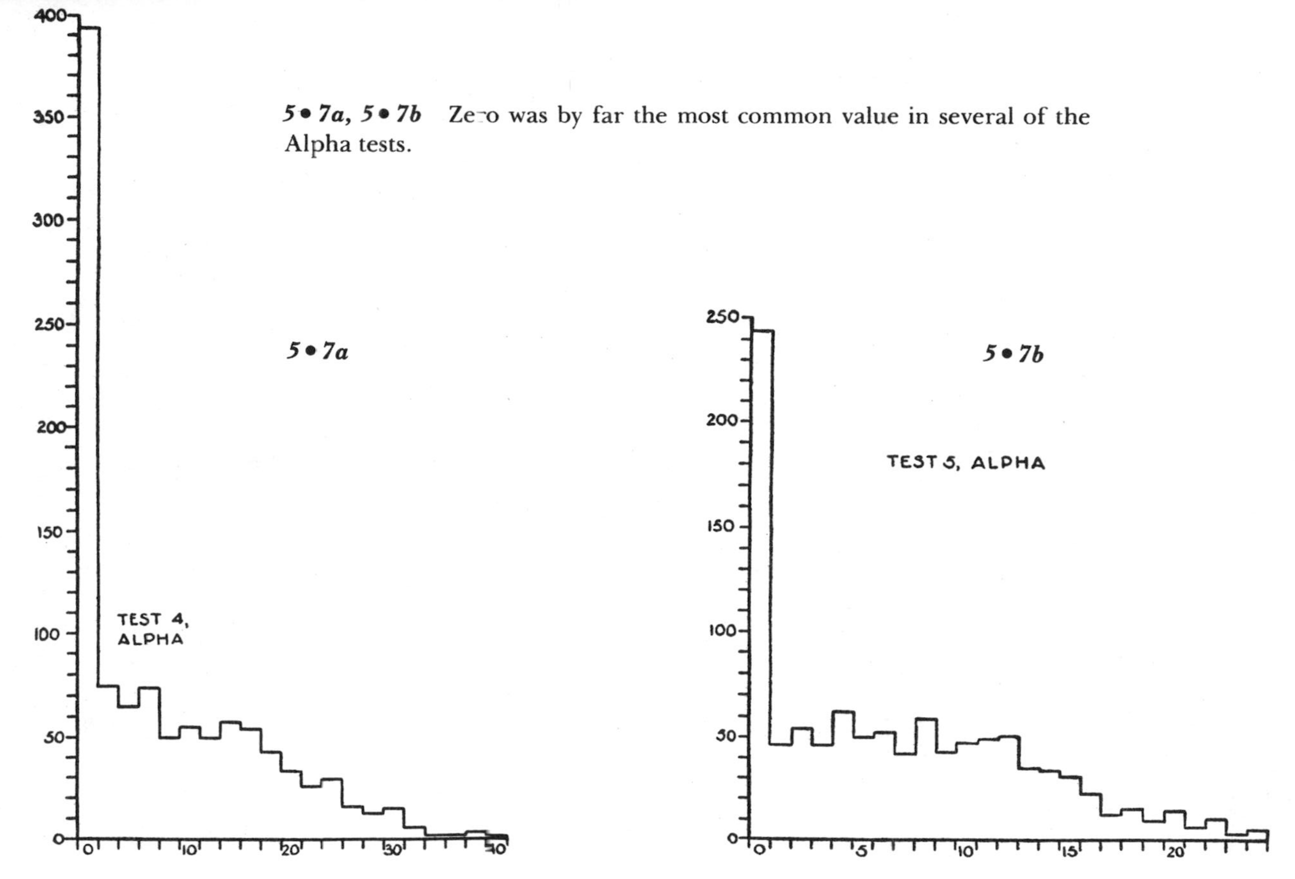

of giving the tests, tended to interpret zeros in this common-sense fashion. They eliminated some tests from the Beta repertoire (p. 372) because they produced up to 30.7 percent zero scores (although some Alpha tests with a higher frequency of zeros were retained). They reduced the difficulty of initial items in several tests "in order to reduce the number of zero scores" (p. 341). They included among the criteria for acceptance of a test within the Beta repertoire (p. 373): "ease of demonstration, as shown by low percentage of zero scores." They acknowledged several times that a high frequency of zeros reflected poor explanation, not stupidity of the recruits: "The large number of zero scores, even with officers, indicates that the instructions were unsatisfactory" (p. 340). "The main burden of the early reports was to the effect that the most difficult task was 'getting the idea across.' A high percentage of zero scores in any given test was considered an indication of failure to 'get that test across' " (p. 379).

With all these acknowledgments, one might have anticipated Boring's decision either to exclude zeros from the summary statistics or to correct for them by assuming that most recruits would have scored some points if they had understood what they were supposed to do. Instead, Boring "corrected" zero scores in the opposite way, and actually demoted many of them into a negative range.

Boring began with the same hereditarian assumption that invalidated all the results: that the tests, by definition, measure innate intelligence. The clump of zeros must therefore be made up of men who were too stupid to do any items. Is it fair to give them all zero? After all, some must have been just barely too stupid, and their zero is a fair score. But other dullards must have been rescued from an even worse fate by the minimum of zero. They would have done even more poorly if the test had included enough easy items to make distinctions among the zero scores. Boring distinguished between a true "mathematical zero," an intrinsic minimum that cannot logically go lower, and a "psychological zero," an arbitrary beginning defined by a particular test. (As a general statement, Boring makes a sound point. In the particular context of the army tests, it is absurd):

> A score of zero, therefore, does not mean no ability at all; it does not mean the point of discontinuance of the thing measured; it means the point of discontinuance of the instrument of measurement, the test. . . . The indi-

vidual who fails to earn a positive score and is marked zero is actually thereby given a bonus varying in value directly with his stupidity (p. 622).

Boring therefore "corrected" each zero score by calibrating it against other tests in the series on which the same man had scored some points. If he had scored well on other tests, he was not doubly penalized for his zeros; if he had done poorly, then his zeros were converted to negative scores.

By this method, a debilitating flaw in Yerkes's basic procedure was accentuated by tacking an additional bias onto it. The zeros only indicated that, for a suite of reasons unrelated to intelligence, vast numbers of men did not understand what they were supposed to do. And Yerkes should have recognized this, for his own reports proved that, with reduced confusion and harassment, men who had scored zero on the group tests almost all managed to make points on the same or similar tests given in an individual examination. He writes (p. 406): "At Greenleaf it was found that the proportion of zero scores in the maze test was reduced from 28 percent in Beta to 2 percent in the performance scale, and that similarly zero scores in the digit-symbol test were reduced from 49 to 6 percent."

Yet, when given an opportunity to correct this bias by ignoring or properly redistributing the zero scores, Yerkes's statisticians did just the opposite. They exacted a double penalty by demoting most zero scores to a negative range.

FINAGLING THE SUMMARY STATISTICS: GETTING AROUND OBVIOUS CORRELATIONS WITH ENVIRONMENT

Yerkes's monograph is a treasure-trove of information for anyone seeking environmental correlates of performance on "tests of intelligence." Since Yerkes explicitly denied any substantial causal role to environment, and continued to insist that the tests measured innate intelligence, this claim may seem paradoxical. One might suspect that Yerkes, in his blindness, didn't read his own information. The situation, in fact, is even more curious. Yerkes read very carefully; he puzzled over every one of his environmental correlations, and managed to explain each of them away with arguments that sometimes border on the ridiculous.

Minor items are reported and dispersed in a page or two. Yerkes found strong correlations between average score and infestation with hookworm in all 4 categories:

	INFECTED	NOT INFECTED
White Alpha	94.38	118.50
White Beta	45.38	53.26
Negro Alpha	34.86	40.82
Negro Beta	22.14	26.09

These results might have led to the obvious admission that state of health, particularly in diseases related to poverty, has some effect upon the scores. Although Yerkes did not deny this possibility, he stressed another explanation (p. 811): "Low native ability may induce such conditions of living as to result in hookworm infection."

In studying the distribution of scores by occupation, Yerkes conjectured that since intelligence brings its own reward, test scores should rise with expertise. He divided each job into apprentices, journeymen, and experts and searched for increasing scores between the groups. But he found no pattern. Instead of abandoning his hypothesis, he decided that his procedure for allocating men to the three categories must have been flawed (pp. 831–832):

> It seems reasonable to suppose that a selection process goes on in industry which results in a selection of the mentally more alert for promotion from the apprentice stage to the journeyman stage and likewise from the journeyman stage to the expert. Those inferior mentally would stick at the lower levels of skill or be weeded out of the particular trade. On this hypothesis one begins to question the accuracy of the personnel interviewing procedure.

Among major patterns, Yerkes continually found relationships between intelligence and amount of schooling. He calculated a correlation coefficient of 0.75 between test score and years of education. Of 348 men who scored below the mean in Alpha, only 1 had ever attended college (as a dental student), 4 had graduated from high school, and only 10 had ever attended high school at all. Yet Yerkes did not conclude that more schooling leads to increasing scores per se; instead, he argued that men with more innate intelligence spend more time in school. "The theory that native intelligence is one of the most important conditioning factors in continuance in school is certainly borne out by this accumulation of data" (p. 780).

Yerkes noted the strongest correlation of scores with schooling in considering the differences between blacks and whites. He made a significant social observation, but gave it his usual innatist twist (p. 760):

> The white draft of foreign birth is less schooled; more than half of this group have not gone beyond the fifth grade, while one-eighth, or 12.5 percent, report no schooling. Negro recruits though brought up in this country where elementary education is supposedly not only free but compulsory on all, report no schooling in astonishingly large proportion.

Failure of blacks to attend school, he argued, must reflect a disinclination based on low innate intelligence. Not a word about segregation (then officially sanctioned, if not mandated), poor conditions in black schools, or economic necessities for working among the impoverished. Yerkes acknowledged that schools might vary in quality, but he assumed that such an effect must be small and cited, as primary evidence for innate black stupidity, the lower scores of blacks when paired with whites who had spent an equal number of years in school (p. 773):

> The grade standards, of course, are not identical all over the country, especially as between schools for white and for negro children, so that "fourth-grade schooling" doubtless varies in meaning from group to group, but this variability certainly cannot account for the clear intelligence differences between groups.

The data that might have led Yerkes to change his mind (had he approached the study with any flexibility) lay tabulated, but unused, within his monograph. Yerkes had noted regional differences in black education. Half the black recruits from Southern states had not attended school beyond the third grade, but half had reached the fifth grade in Northern states (p. 760). In the North, 25 percent completed primary school; in the South, a mere 7 percent. Yerkes also noted (p. 734) that "the percentage of Alphas is very much smaller and the percentage of Betas very much larger in the southern than in the northern group." Many years later, Ashley Montagu (1945) studied the tabulations by state that Yerkes had provided. He confirmed Yerkes's pattern: the average score on Alpha was 21.31 for blacks in thirteen Southern states, and 39.90 in nine Northern states. Montagu then noted that average black scores for the four highest Northern states (45.31) exceeded the *white* mean for nine Southern states (43.94). He found the same pattern for Beta, where blacks of six Northern states averaged 34.63, and whites of fourteen Southern states, 31.11. Hereditarians had their pat answer, as usual: only the best Negroes had been smart enough to move North. To people of good will and common sense an explanation in terms of educational quality has always seemed more reasonable, especially since Montagu also found such high correlations between a state's expenditure for education and the average score of its recruits.

One other persistent correlation threatened Yerkes's hereditarian convictions, and his rescuing argument became a major social weapon in later political campaigns for restricting immigration. Test scores had been tabulated by country of origin, and Yerkes noted the pattern so dear to the hearts of Nordic supremacists. He divided recruits by country of origin into English, Scandinavian, and Teutonic on one side, and Latin and Slavic on the other, and stated (p. 699): "the differences are considerable (an extreme range of practically two years mental age)"—favoring the Nordics, of course.

But Yerkes acknowledged a potential problem. Most Latins and Slavs had arrived recently and spoke English either poorly or not at all; the main wave of Teutonic immigration had passed long before. According to Yerkes's protocol, it shouldn't have mattered. Men who could not speak English suffered no penalty. They took Beta, a pictorial test that supposedly measured innate ability independent of literacy and language. Yet the data still showed an apparent penalty for unfamiliarity with English. Of white recruits who scored E in Alpha and therefore took Beta as well (pp. 382–383), speakers of English averaged 101.6 in Beta, while nonspeakers averaged only 77.8. On the individual performance scale, which eliminated the harassment and confusion of Beta, native and foreign-born recruits did not differ (p. 403). (But very few men were ever given these individual tests, and they did not affect national averages.) Yerkes had to admit (p. 395): "There are indications to the effect that individuals handicapped by language difficulty and illiteracy are penalized to an appreciable degree in Beta as compared with men not so handicapped."

Another correlation was even more potentially disturbing. Yerkes found that average test scores for foreign-born recruits rose consistently with years of residence in America.

YEARS OF RESIDENCE	AVERAGE MENTAL AGE
0–5	11.29
6–10	11.70
11–15	12.53
16–20	13.50
20–	13.74

Didn't this indicate that familiarity with American ways, and not innate intelligence, regulated the differences in scores? Yerkes admitted the possibility, but held out strong hope for a hereditarian salvation (p. 704):

> Apparently then the group that has been longer resident in this country does somewhat better* in intelligence examination. It is not possible to state whether the difference is caused by the better adaptation of the more thoroughly Americanized group to the situation of the examination or whether some other factor is operative. It might be, for instance, that the more intelligent immigrants succeed and therefore remain in this country, but this suggestion is weakened by the fact that so many successful immigrants do return to Europe. At best we can but leave for future decision the question as to whether the differences represent a real difference of intelligence or an artifact of the method of examination.

The Teutonic supremacists would soon supply that decision: recent immigration had drawn the dregs of Europe, lower-class Latins and Slavs. Immigrants of longer residence belonged predominantly to superior northern stocks. The correlation with years in America was an artifact of genetic status.

The army mental tests could have provided an impetus for social reform, since they documented that environmental disadvantages were robbing from millions of people an opportunity to develop their intellectual skills. Again and again, the data pointed to strong correlations between test scores and environment. Again and again, those who wrote and administered the tests invented tortuous, ad hoc explanations to preserve their hereditarian prejudices.

How powerful the hereditarian biases of Terman, Goddard, and Yerkes must have been to make them so blind to immediate

*Note how choice of language can serve as an indication of bias. This 2.5 year difference in mental ages (13.74–11.29) only represents "somewhat better" performance. The smaller (but presumably hereditary) difference of 2 years between Nordic-Teutonic and Latin–Slav groups had been described as "considerable."

circumstances! Terman seriously argued that good orphanages precluded any environmental cause of low IQ for children in them. Goddard tested confused and frightened immigrants who had just completed a grueling journey in steerage and thought he had captured innate intelligence. Yerkes badgered his recruits, obtained proof of confusion and harassment in their large mode of zero scores, and produced data on the inherent abilities of racial and national groups. One cannot attribute all these conclusions to some mysterious "temper of the times," for contemporary critics saw through the nonsense as well. Even by standards of their own era, the American hereditarians were dogmatists. But their dogma wafted up on favorable currents into realms of general acceptance, with tragic consequences.

Political impact of the army data

CAN DEMOCRACY SURVIVE AN AVERAGE MENTAL AGE OF THIRTEEN?

Yerkes was troubled by his own figure of 13.08 as an average mental age for the white draft. It fitted his prejudices and the eugenical fears of prosperous old Americans, but it was too good to be true, or too low to be believed. Yerkes recognized that smarter folks had been excluded from the sample—officers who enlisted and "professional and business experts that were exempted from draft because essential to industrial activity in the war" (p. 785). But the obviously retarded and feeble-minded had also been culled before reaching Yerkes's examiners, thereby balancing exclusions at the other end. The resulting average of 13 might be a bit low, but it could not be far wrong (p. 785).

Yerkes faced two possibilities. He could recognize the figure as absurd, and search his methods for the flaws that engendered such nonsense. He would not have had far to look, had he been so inclined, since three major biases all conspired to bring the average down to his implausible figure. First, the tests measured education and familiarity with American culture, not innate intelligence—and many recruits, whatever their intelligence, were both woefully deficient in education and either too new to America or too impoverished to have much appreciation for the exemplary accomplishments of Mr. Mathewson (including an e.r.a. of 1.14 in 1909). Second, Yerkes's own stated protocol had not been followed. About two-thirds of the white sample took Alpha, and their high frequency of zero scores indicated that many should have been retested in Beta. But time and the indifference of the regular brass conspired against it, and many recruits were not reexamined. Finally, Boring's treatment of zero values imposed an additional penalty on scores already (and artificially) too low.

Or Yerkes could accept the figure and remain a bit puzzled. He opted, of course, for the second strategy:

> We know now approximately from clinical experience the capacity and mental ability of a man of 13 years mental age. We have never heretofore supposed that the mental ability of this man was the average of the country or anywhere near it. A moron has been defined as anyone with a mental age from 7 to 12 years. If this definition is interpreted as meaning anyone with a mental age less than 13 years, as has recently been done, then almost half of the white draft (47.3 percent) would have been morons. Thus it appears that feeble-mindedness, as at present defined, is of much greater frequency of occurrence than had been originally supposed.

Yerkes's colleagues were disturbed as well. Goddard, who had invented the moron, began to doubt his own creation: "We seem to be impaled on the horns of a dilemma: either half the population is feeble-minded; or 12 year mentality does not properly come within the limits of feeble-mindedness" (1919, p. 352). He also opted for Yerkes's solution and sounded the warning cry for American democracy:

> If it is ultimately found that the intelligence of the average man is 13—instead of 16—it will only confirm what some are beginning to suspect; viz., that the average man can manage his affairs with only a moderate degree of prudence, can earn only a very modest living, and is vastly better off when following directions than when trying to plan for himself. In other words, it will show that there is a fundamental reason for many of the conditions that we find in human society and further that much of our effort to change conditions is unintelligent because we have not understood the nature of the average man (1919, p. 236).

Unfortunate 13 became a formula figure among those who sought to contain movements for social welfare. After all, if the average man is scarcely better than a moron, then poverty is fundamentally biological in origin, and neither education nor better opportunities for employment can alleviate it. In a famous address, entitled "Is America safe for democracy?", the chairman of Harvard's psychology department stated (W. McDougall, quoted in Chase, 1977, p. 226):

> The results of the Army tests indicate that about 75 percent of the population has not sufficient innate capacity for intellectual development to enable it to complete the usual high school course. The very extensive testing of school-children carried on by Professor Terman and his colleagues leads to closely concordant results.

In an inaugural address as president of Colgate University, G. G. Cutten proclaimed in 1922 (quoted in Cravens, 1978, p. 224): "We cannot conceive of any worse form of chaos than a real democracy in a population of average intelligence of a little over 13 years."

Again, a catchy, numerical "fact" had risen to prominence as the discovery of objective science—while the fallacies and finagling that thoroughly invalidated it remained hidden in the details of an eight-hundred-page monograph that the propagandists never read.

THE ARMY TESTS AND AGITATION TO RESTRICT IMMIGRATION: BRIGHAM'S MONOGRAPH ON AMERICAN INTELLIGENCE

The grand average of thirteen had political impact, but its potential for social havoc was small compared with Yerkes's figures for racial and national differences; for hereditarians could now claim that the fact and extent of group differences in innate intelligence had finally, once and for all, been established. Yerkes's disciple C. C. Brigham, then an assistant professor of psychology at Princeton University, proclaimed (1923, p. xx):

> We have here an investigation which, of course, surpasses in reliability all preceding investigations, assembled and correlated, a hundred fold. These army data constitute the first really significant contribution to the study of race differences in mental traits. They give us a scientific basis for our conclusions.

In 1923 Brigham published a book, short enough and stated with sufficient baldness (some would say clarity) to be read and used by all propagandists. *A Study of American Intelligence* (Brigham, 1923) became a primary vehicle for translating the army results on group differences into social action (see Kamin, 1974 and Chase, 1977). Yerkes himself wrote the foreword and praised Brigham for his objectivity:

> The author presents not theories or opinion but facts. It behooves us to consider their reliability and their meaning, for no one of us as a citizen can afford to ignore the menace of race deterioration or the evident rela-

tions of immigration to national progress and welfare (in Brigham, 1923, p. vii).

Since Brigham derived his "facts" on group differences entirely from the army results, he had first to dismiss the claim that Yerkes's tests might not be pure measures of innate intelligence. He admitted that Alpha might mingle the impact of education with native ability, for it did require literacy. But Beta could only record unadulterated innate intelligence: "Examination Beta involves no English, and the tests cannot be considered as educational measures in any sense" (p. 100). In any case, he added for good measure, it scarcely matters whether the tests also record what Yerkes had called "the better adaptation of the more thoroughly Americanized group to the situation of the examination" (p. 93), since (p. 96):

If the tests used included some mysterious type of situation that was "typically American," we are indeed fortunate, for this is America, and the purpose of our inquiry is that of obtaining a measure of the character of our immigration.* Inability to respond to a "typically American" situation is obviously an undesirable trait.

Once he had proved that the tests measure innate intelligence, Brigham devoted most of his book to dispelling common impressions that might threaten this basic assumption. The army tests had, for example, assessed Jews (primarily recent immigrants) as quite low in intelligence. Does this discovery not conflict with the notable accomplishments of so many Jewish scholars, statesmen, and performing artists? Brigham conjectured that Jews might be more variable than other groups; a low mean would not preclude a few geniuses in the upper range. In any case, Brigham added, we probably focus unduly on the Jewish heritage of some great men because it surprises us: "The able Jew is popularly recognized not only because of his ability, but because he is able and a Jew" (p. 190). "Our figures, then, would rather tend to disprove the popular belief that the Jew is highly intelligent" (p. 190).

But what about the higher scores of Northern vs. Southern blacks? Since Yerkes had also shown that Northern blacks, on average, attended school for several more years than their Southern counterparts, didn't the scores reflect differences in education

*In all other parts of the book, he claims that his aim is to measure and interpret innate differences in intelligence.

more than inborn ability? Brigham did not deny a small effect for education (p. 191), but he presented two reasons for attributing the higher scores of Northern blacks primarily to better biology: first, "the greater admixture of white blood" among Northern blacks; second, "the operation of economic and social forces, such as higher wages, better living conditions, identical school privileges, and a less complete social ostracism, tending to draw the more intelligent negro to the north" (p. 192).

Brigham faced the greatest challenge to hereditarianism on the issue of immigration. Even Yerkes had expressed agnosticism—the only time he considered a significant alternative to inborn biology—on the causes of steadily increasing scores for immigrants who had lived longer in America (see p. 221). The effects were certainly large, the regularity striking. Without exception (see chart on p. 221), each five years of residency brought an increase in test scores, and the total difference between recent arrivals and the longest residents was a full two and a half years in mental age.

Brigham directed himself around the appalling possibility of environmentalism by arguing in a circle. He began by assuming what he intended to demonstrate. He denied the possibility of environmental influence a priori, by accepting as proven the highly controversial claim that Beta must measure unadulterated innate intelligence, whatever Alpha may be doing with its requirement of literacy. The biological basis of declining scores for recent immigrants can then be proven by demonstrating that decrease on the combined scale is not an artifact of differences in Alpha only:

> The hypothesis of growth of intelligence with increasing length of residence may be identified with the hypothesis of an error in the method of measuring intelligence, for we must assume that we are measuring native or inborn intelligence, and any increase in our test score due to any other factor may be regarded as an error. . . . If all members of our five years of residence groups had been given Alpha, Beta, and individual examinations in equal proportions, then all would have been treated alike, and the relationship shown would stand without any possibility of error (p. 100).

If the differences between residence groups are not innate, Brigham argued, then they reflect a technical flaw in constructing the combined scale from varying proportions of Alphas and Betas; they cannot arise from a defect in the tests themselves, and therefore cannot, by definition, be environmental indicators of increasing familiarity with American customs and language.

Brigham studied the performances of Alphas and Betas, found that differences between residence groups persisted among the Betas, and proclaimed his counter-intuitive hypothesis of decreasing innate intelligence among more recent immigrants. "We actually find," he proclaimed (p. 102), "that the gain from each type of examination [both Alpha and Beta] is about the same. This indicates, then, that the five years of residence groups are groups with real differences in native intelligence, and not groups laboring under more or less of a linguistic and educational handicap."

> Instead of considering that our curve indicates a growth of intelligence with increasing length of residence, we are forced to take the reverse of the picture and accept the hypothesis that the curve indicates a gradual deterioration in the class of immigrants examined in the army, who came to this country in each succeeding 5 year period since 1902 (pp. 110–111). . . . The average intelligence of succeeding waves of immigration has become progressively lower (p. 155).

But why should recent immigrants be more stupid? To resolve this conundrum, Brigham invoked the leading theorist of racism in his day, the American Madison Grant (author of *The Passing of the Great Race*), and that aging relic from the heyday of French craniometry, Count Georges Vacher de Lapouge. Brigham argued that the European peoples are mixtures, to varying degrees, of three original races: 1) Nordics, "a race of soldiers, sailors, adventurers, and explorers, but above all, of rulers, organizers, and aristocrats . . . feudalism, class distinctions, and race pride among Europeans are traceable for the most part to the North." They are "domineering, individualistic, self-reliant . . . and as a result they are usually Protestants" (Grant, quoted in Brigham, p. 182); 2) Alpines, who are "submissive to authority both political and religious, being usually Roman Catholics" (Grant, in Brigham, p. 183), and whom Vacher de Lapouge described as "the perfect slave, the ideal serf, the model subject" (p. 183); 3) Mediterraneans, of whom Grant approved, given their accomplishments in ancient Greece and Rome, but whom Brigham despised because their average scores were even slightly lower than the Alpines.

Brigham then tried to assess the amount of Nordic, Alpine, and Mediterranean blood in various European peoples, and to calculate the army scores on this scientific and racial basis, rather than from the political expedient of national origin. He devised the following figures for average intelligence: Nordic, 13.28; Alpine, 11.67; Mediterranean, 11.43.

The progressive decline of intelligence for each five-year residency group then achieved its easy, innatist explanation. The character of immigration had changed markedly during the past twenty years. Before then, arrivals had been predominantly Nordic; since then, we have been inundated by a progressively increasing number of Alpines and Mediterraneans, as the focus of immigration shifted from Germany, Scandinavia, and the British Isles to the great unwashed of southern and eastern Europe—Italians, Greeks, Turks, Hungarians, Poles, Russians, and other Slavs (including Jews, whom Brigham defined racially as "Alpine Slavs"). Of the inferiority of these recent immigrants, there can be no doubt (p. 202):

> The Fourth of July orator can convincingly raise the popular belief in the intellectual level of Poland by shouting the name of Kosciusko from a high platform, but he cannot alter the distribution of the intelligence of the Polish immigrant.

But Brigham realized that two difficulties still stood before his innatist claim. He had proved that the army tests measured inborn intelligence, but he still feared that ignorant opponents might try to attribute high Nordic scores to the presence of so many native speakers of English in the group.

He therefore divided the Nordic group into native speakers from Canada and the British isles, who averaged 13.84, and "non-English speakers," primarily from Germany, Holland, and Scandinavia, who averaged 12.97. Again, Brigham had virtually proved the environmentalist claim that army tests measured familiarity with American language and customs; but again, he devised an innatist fudge. The disparity between English and non-English Nordics was half as large as the difference between Nordics and Mediterraneans. Since differences among Nordics could only represent the environmental effects of language and culture (as Brigham admitted), why not attribute variation between European races to the same cause? After all, the so-called non-English Nordics were, on average, more familiar with American ways and should have scored higher than Alpines and Mediterraneans on this basis alone. Brigham called these men "non-English" and used them as a test of his language hypothesis. But, in fact, he only knew their country of origin, not their degree of familiarity with English. On average, these so-called non-English Nordics had been in America far longer than the Alpines or Mediterraneans. Many spoke English well and had spent enough years in America to master the arcana of bowling, commercial products, and film stars. If they, with their intermediary knowledge of American culture, scored almost a year below the English Nordics, why not attribute the nearly two-year disadvantage of Alpines and Mediterraneans to their greater average unfamiliarity with American ways? It is surely more parsimonious to use the same explanation for a continuum of effects. Instead, Brigham admitted environmental causes for the disparity within Nordics, but then advanced innatism to explain the lower scores of his despised southern and eastern Europeans (pp. 171–172):

> There are, of course, cogent historical and sociological reasons accounting for the inferiority of the non-English speaking Nordic group. On the other hand, if one wishes to deny, in the teeth of the facts, the superiority of the Nordic race on the ground that the language factor mysteriously aids this group when tested, he may cut out of the Nordic distribution the English speaking Nordics, and still find a marked superiority of the non-English speaking Nordics over the Alpine and Mediterranean groups, a fact which clearly indicates that the underlying cause of the nativity differences we have shown is race, and not language.

Having met this challenge, Brigham encountered another that he couldn't quite encompass. He had attributed the declining scores of successive five-year groups to the decreasing percentage of Nordics in their midst. Yet he had to admit a troubling anachronism. The Nordic wave had diminished long before, and immigration for the two or three most recent five-year groups had included a roughly constant proportion of Alpines and Mediterraneans. Yet scores continued to drop while racial composition remained constant. Didn't this, at least, implicate language and culture? After all, Brigham had avoided biology in explaining the substantial differences between Nordic groups; why not treat similar differences among Alpines and Mediterraneans in the same way? Again, prejudice annihilated common sense and Brigham invented an implausible explanation for which, he admitted, he had no direct evidence. Since scores of Alpines and Mediterraneans had been declining, the nations harboring these miscreants must be sending a progressively poorer biological stock as the years wear on (p. 178):

> The decline in intelligence is due to two factors, the change in the races migrating to this country, and to the additional factor of the sending of lower and lower representatives of each race.

The prospects for America, Brigham groused, were dismal. The European menace was bad enough, but America faced a special and more serious problem (p. xxi):

> Running parallel with the movements of these European peoples, we have the most sinister development in the history of this continent, the importation of the negro.

Brigham concluded his tract with a political plea, advocating the hereditarian line on two hot political subjects of his time: the restriction of immigration and eugenical regulation of reproduction (pp. 209–210):

> The decline of American intelligence will be more rapid than the decline of the intelligence of European national groups, owing to the presence here of the negro. These are the plain, if somewhat ugly, facts that our study shows. The deterioration of American intelligence is not inevitable, however, if public action can be aroused to prevent it. There is no reason why legal steps should not be taken which would insure a continuously progressive upward evolution.
>
> The steps that should be taken to preserve or increase our present intellectual capacity must of course be dictated by science and not by political expediency. Immigration should not only be restrictive but highly selective. And the revision of the immigration and naturalization laws will only afford a slight relief from our present difficulty. The really important steps are those looking toward the prevention of the continued propagation of defective strains in the present population.

As Yerkes had said of Brigham: "The author presents not theories or opinions but facts."

THE TRIUMPH OF RESTRICTION ON IMMIGRATION

The army tests engendered a variety of social uses. Their most enduring effect surely lay in the field of mental testing itself. They were the first written IQ tests to gain respect, and they provided essential technology for implementing the hereditarian ideology that advocated, contrary to Binet's wishes, the testing and ranking of all children.

Other propagandists used the army results to defend racial segregation and limited access of blacks to higher education. Cornelia James Cannon, writing in the *Atlantic Monthly* in 1922, noted that 89 percent of blacks had tested as morons and argued (quoted in Chase, 1977, p. 263):

> Emphasis must necessarily be laid on the development of the primary schools, on the training in activities, habits, occupations which do not demand the more evolved faculties. In the South particularly . . . the education of the whites and colored in separate schools may have justification other than that created by race prejudice. . . . A public school system, preparing for life young people of a race, 50 percent of whom never reach a mental age of 10, is a system yet to be perfected.

But the army data had its most immediate and profound impact upon the great immigration debate, then a major political issue in America, and ultimately the greatest triumph of eugenics. Restriction was in the air, and may well have occurred without scientific backing. (Consider the wide spectrum of support that limitationists could muster—from traditional craft unions fearing multitudes of low-paid laborers, to jingoists and America firsters who regarded most immigrants as bomb-throwing anarchists and who helped make martyrs of Sacco and Vanzetti.) But the timing, and especially the peculiar character, of the 1924 Restriction Act clearly reflected the lobbying of scientists and eugenicists, and the army data formed their most powerful battering ram (see Chase, 1977; Kamin, 1974; and Ludmerer, 1972).

Henry Fairfield Osborn, trustee of Columbia University and president of the American Museum of Natural History, wrote in 1923, in a statement that I cannot read without a shudder when I recall the gruesome statistics of mortality for World War I:

> I believe those tests were worth what the war cost, even in human life, if they served to show clearly to our people the lack of intelligence in our country, and the degrees of intelligence in different races who are coming to us, in a way which no one can say is the result of prejudice. . . . We have learned once and for all that the negro is not like us. So in regard to many races and subraces in Europe we learned that some which we had believed possessed of an order of intelligence perhaps superior to ours [read Jews] were far inferior.

Congressional debates leading to passage of the Immigration Restriction Act of 1924 continually invoke the army data. Eugenicists lobbied not only for limits to immigration, but for changing its character by imposing harsh quotas against nations of inferior stock—a feature of the 1924 act that might never have been implemented, or even considered, without the army data and eugenicist propaganda. In short, southern and eastern Europeans, the Alpine and Mediterranean nations with minimal scores on the army tests, should be kept out. The eugenicists battled and won one of the greatest victories of scientific racism in American history. The first restriction act of 1921 had set yearly quotas at 3 percent of immigrants from any nation then resident in America. The 1924 act, following a barrage of eugenicist propaganda, reset the quotas at 2 percent of people from each nation recorded in the 1890 census. The 1890 figures were used until 1930. Why 1890 and not 1920 since the act was passed in 1924? 1890 marked a watershed in the history of immigration. Southern and eastern Europeans arrived in relatively small numbers before then, but began to predominate thereafter. Cynical, but effective. "America must be kept American," proclaimed Calvin Coolidge as he signed the bill.

BRIGHAM RECANTS

Six years after his data had so materially affected the establishment of national quotas, Brigham had a profound change of heart. He recognized that a test score could not be reified as an entity inside a person's head:

> Most psychologists working in the test field have been guilty of a naming fallacy which easily enables them to slide mysteriously from the score in the test to the hypothetical faculty suggested by the name given to the test. Thus, they speak of sensory discrimination, perception, memory, intelligence, and the like while the reference is to a certain objective test situation (Brigham, 1930, p. 159).

In addition, Brigham now realized that the army data were worthless as measures of innate intelligence for two reasons. For each error, he apologized with an abjectness rarely encountered in scientific literature. First, he admitted that Alpha and Beta could not be combined into a single scale as he and Yerkes had done in producing averages for races and nations. The tests measured different things, and each was internally inconsistent in any case. Each nation was represented by a sample of recruits who had taken Alpha and Beta in differing proportions. Nations could not be compared at all (Brigham, 1930, p. 164):

> As this method of amalgamating Alphas and Betas to produce a combined scale was used by the writer in his earlier analysis of the Army tests as applied to samples of foreign born in the draft, that study with its entire hypothetical superstructure of racial differences collapses completely.

Secondly, Brigham acknowledged that the tests had measured familiarity with American language and culture, not innate intelligence:

> For purposes of comparing individuals or groups, it is apparent that tests in the vernacular must be used only with individuals having equal opportunity to acquire the vernacular of the test. This requirement precludes the use of such tests in making comparative studies of individuals brought up in homes in which the vernacular of the test is not used, or in which two vernaculars are used. The last condition is frequently violated here in studies of children born in this country whose parents speak another tongue. It is important, as the effects of bilingualism are not entirely known. . . . Comparative studies of various national and racial groups may not be made with existing tests. . . . One of the most pretentious of these comparative racial studies—the writer's own—was without foundation (Brigham, 1930, p. 165).

Brigham paid his personal debt, but he could not undo what the tests had accomplished. The quotas stood, and slowed immigration from southern and eastern Europe to a trickle. Throughout the 1930s, Jewish refugees, anticipating the holocaust, sought to emigrate, but were not admitted. The legal quotas, and continuing eugenical propaganda, barred them even in years when inflated quotas for western and northern European nations were not filled. Chase (1977) has estimated that the quotas barred up to 6 million southern, central, and eastern Europeans between 1924 and the outbreak of World War II (assuming that immigration had continued at its pre-1924 rate). We know what happened to many who wished to leave but had nowhere to go. The paths to destruction are often indirect, but ideas can be agents as sure as guns and bombs.

SIX

The Real Error of Cyril Burt

Factor Analysis and the Reification of Intelligence

It has been the signal merit of the English school of psychology, from Sir Francis Galton onwards, that it has, by this very device of mathematical analysis, transformed the mental test from a discredited dodge of the charlatan into a recognized instrument of scientific precision.

—CYRIL BURT, 1921, p. 130

The case of Sir Cyril Burt

If I had any desire to lead a life of indolent ease, I would wish to be an identical twin, separated at birth from my brother and raised in a different social class. We could hire ourselves out to a host of social scientists and practically name our fee. For we would be exceedingly rare representatives of the only really adequate natural experiment for separating genetic from environmental effects in humans—genetically identical individuals raised in disparate environments.

Studies of identical twins raised apart should therefore hold pride of place in literature on the inheritance of IQ. And so it would be but for one problem—the extreme rarity of the animal itself. Few investigators have been able to rustle up more than twenty pairs of twins. Yet, amidst this paltriness, one study seemed to stand out: that of Sir Cyril Burt (1883–1971). Sir Cyril, doyen of mental testers, had pursued two sequential careers that gained him a preeminent role in directing both theory and practice in his field of educational psychology. For twenty years he was the official psychologist of the London County Council, responsible for the administration and interpretation of mental tests in London's schools. He then succeeded Charles Spearman as professor in the most influential chair of psychology in Britain: University College, London (1932–1950). During his long retirement, Sir Cyril published several papers that buttressed the hereditarian claim by citing very high correlation between IQ scores of identical twins raised apart. Burt's study stood out among all others because he had found fifty-three pairs, more than twice the total of any previous attempt. It is scarcely surprising that Arthur Jensen used Sir Cyril's figures as the most important datum in his notorious article (1969) on supposedly inherited and ineradicable differences in intelligence between whites and blacks in America.

The story of Burt's undoing is now more than a twice-told tale. Princeton psychologist Leon Kamin first noted that, while Burt had increased his sample of twins from fewer than twenty to more than fifty in a series of publications, the average correlation between pairs for IQ remained unchanged to the third decimal place—a statistical situation so unlikely that it matches our vernacular definition of impossible. Then, in 1976, Oliver Gillie, medical correspondent of the London *Sunday Times,* elevated the charge from inexcusable carelessness to conscious fakery. Gillie discovered, among many other things, that Burt's two "collaborators," a Margaret Howard and a J. Conway, the women who supposedly collected and processed his data, either never existed at all, or at least could not have been in contact with Burt while he wrote the papers bearing their names. These charges led to further reassessments of Burt's "evidence" for his rigid hereditarian position. Indeed, other crucial studies were equally fraudulent, particularly his IQ correlations between close relatives (suspiciously too good to be true and apparently constructed from ideal statistical distributions, rather than measured in nature—Dorfman, 1978), and his data for declining levels of intelligence in Britain.

Burt's supporters tended at first to view the charges as a thinly veiled leftist plot to undo the hereditarian position by rhetoric. H. J. Eysenck wrote to Burt's sister: "I think the whole affair is just a determined effort on the part of some very left-wing environmentalists determined to play a political game with scientific facts. I am sure the future will uphold the honor and integrity of Sir Cyril without any question." Arthur Jensen, who had called Burt a "born nobleman" and "one of the world's great psychologists," had to conclude that the data on identical twins could not be trusted, though he attributed their inaccuracy to carelessness alone.

I think that the splendid "official" biography of Burt recently published by L. S. Hearnshaw (1979) has resolved the issue so far as the data permit (Hearnshaw was commissioned to write his book by Burt's sister before any charges had been leveled). Hearnshaw, who began as an unqualified admirer of Burt and who tends to share his intellectual attitudes, eventually concluded that all allegations are true, and worse. And yet, Hearnshaw has convinced me that the very enormity and bizarreness of Burt's fakery forces us to view it not as the "rational" program of a devious person trying to salvage his hereditarian dogma when he knew the game was up (my original suspicion, I confess), but as the actions of a sick and tortured man. (All this, of course, does not touch the deeper issue of why such patently manufactured data went unchallenged for so long, and what this will to believe implies about the basis of our hereditarian presuppositions.)

Hearnshaw believes that Burt began his fabrications in the early 1940s, and that his earlier work was honest, though marred by rigid a priori conviction and often inexcusably sloppy and superficial, even by the standards of his own time. Burt's world began to collapse during the war, partly by his own doing to be sure. His research data perished in the blitz of London; his marriage failed; he was excluded from his own department when he refused to retire gracefully at the mandatory age and attempted to retain control; he was removed as editor of the journal he had founded, again after declining to cede control at the specified time he himself had set; his hereditarian dogma no longer matched the spirit of an age that had just witnessed the holocaust. In addition, Burt apparently suffered from Ménières disease, a disorder of the organs of balance, with frequent and negative consequences for personality as well.

Hearnshaw cites four instances of fraud in Burt's later career. Three I have already mentioned (fabrication of data on identical twins, kinship correlations in IQ, and declining levels of intelligence in Britain). The fourth is, in many ways, the most bizarre tale of all because Burt's claim was so absurd and his actions so patent and easy to uncover. It could not have been the act of a rational man. Burt attempted to commit an act of intellectual parricide by declaring himself, rather than his predecessor and mentor Charles Spearman, as the father of a technique called "factor analysis" in psychology. Spearman had essentially invented the technique in a celebrated paper of 1904. Burt never challenged this priority—in fact he constantly affirmed it—while Spearman held the chair that Burt would later occupy at University College. Indeed, in his famous book on factor analysis (1940), Burt states that "Spearman's preeminence is acknowledged by every factorist" (1940, p. x).

Burt's first attempt to rewrite history occurred while Spearman was still alive, and it elicited a sharp rejoinder from the occupant emeritus of Burt's chair. Burt withdrew immediately and wrote a letter to Spearman that may be unmatched for deference and obsequiousness: "Surely you have a prior claim here. . . . I have been wondering where precisely I have gone astray. Would it be simplest for me to number my statements, then like my schoolmaster of old you can put a cross against the points where your pupil has blundered, and a tick where your view is correctly interpreted."

But when Spearman died, Burt launched a campaign that "became increasingly unrestrained, obsessive and extravagant" (Hearnshaw, 1979) throughout the rest of his life. Hearnshaw notes (1979, pp. 286–287): "The whisperings against Spearman that were just audible in the late 1930's swelled into a strident campaign of belittlement, which grew until Burt arrogated to himself the whole of Spearman's fame. Indeed, Burt seemed to be becoming increasingly obsessed with questions of priority, and increasingly touchy and egotistical." Burt's false story was simple enough: Karl Pearson had invented the technique of factor analysis (or something close enough to it) in 1901, three years before Spearman's paper. But Pearson had not applied it to psychological problems. Burt recognized its implications and brought the technique into studies of mental testing, making several crucial modifications and improvements along the way. The line, therefore, runs from Pearson to Burt. Spearman's 1904 paper was merely a diversion.

Burt told his story again and again. He even told it through one of his many aliases in a letter he wrote to his own journal and signed Jacques Lafitte, an unknown French psychologist. With the exception of Voltaire and Binet, M. Lafitte cited only English sources and stated: "Surely the first formal and adequate statement was Karl Pearson's demonstration of the method of principal axes in 1901." Yet anyone could have exposed Burt's story as fiction after an hour's effort—for Burt never cited Pearson's paper in any of his work before 1947, while all his earlier studies of factor analysis grant credit to Spearman and clearly display the derivative character of Burt's methods.

Factor analysis must have been very important if Burt chose to center his quest for fame upon a rewrite of history that would make him its inventor. Yet, despite all the popular literature on IQ in the history of mental testing, virtually nothing has been written (outside professional circles) on the role, impact, and meaning of factor analysis. I suspect that the main reason for this neglect lies in the abstrusely mathematical nature of the technique. IQ, a linear scale first established as a rough, empirical measure, is easy to understand. Factor analysis, rooted in abstract statistical theory and based on the attempt to discover "underlying" structure in large matrices of data, is, to put it bluntly, a bitch. Yet this inattention to factor analysis is a serious omission for anyone who wishes to understand the history of mental testing in our century, and its continuing rationale today. For as Burt correctly noted (1914, p. 36), the history of mental testing contains two major and related strands: age-scale methods (Binet IQ testing), and correlational methods (factor analysis). Moreover, as Spearman continually stressed throughout his career, the theoretical justification for using a unilinear scale of IQ resides in factor analysis itself. Burt may have been perverse in his campaign, but he was right in his chosen tactic—a permanent and exalted niche in the pantheon of psychology lies reserved for the man who developed factor analysis.

I began my career in biology by using factor analysis to study the evolution of a group of fossil reptiles. I was taught the technique as though it had developed from first principles using pure logic. In fact, virtually all its procedures arose as justifications for particular theories of intelligence. Factor analysis, despite its status as pure deductive mathematics, was invented in a social context, and for definite reasons. And, though its mathematical basis is unassailable, its persistent use as a device for learning about the physical structure of intellect has been mired in deep conceptual errors from the start. The principal error, in fact, has involved a major theme of this book: reification—in this case, the notion that such a nebulous, socially defined concept as intelligence might be identified as a "thing" with a locus in the brain and a definite degree of heritability—and that it might be measured as a single number, thus permitting a unilinear ranking of people according to the amount of it they possess. By identifying a mathematical factor axis with a concept of "general intelligence," Spearman and Burt provided a theoretical justification for the unilinear scale that Binet had proposed as a rough empirical guide.

The intense debate about Cyril Burt's work has focused exclusively on the fakery of his late career. This perspective has clouded Sir Cyril's greater influence as the most powerful mental tester committed to a factor-analytic model of intelligence as a real and unitary "thing." Burt's commitment was rooted in the error of reification. Later fakery was the afterthought of a defeated man; his earlier, "honest" error has reverberated throughout our century and has affected millions of lives.

Correlation, cause, and factor analysis

Correlation and cause

The spirit of Plato dies hard. We have been unable to escape the philosophical tradition that what we can see and measure in the world is merely the superficial and imperfect representation of an underlying reality. Much of the fascination of statistics lies embedded in our gut feeling—and never trust a gut feeling—that abstract measures summarizing large tables of data must express something more real and fundamental than the data themselves. (Much professional training in statistics involves a conscious effort to counteract this gut feeling.) The technique of *correlation* has been particularly subject to such misuse because it seems to provide a path for inferences about causality (and indeed it does, sometimes—but only sometimes).

Correlation assesses the tendency of one measure to vary in concert with another. As a child grows, for example, both its arms and legs get longer; this joint tendency to change in the same direction is called a *positive correlation.* Not all parts of the body display such positive correlations during growth. Teeth, for example, do not grow after they erupt. The relationship between first incisor length and leg length from, say, age ten to adulthood would represent *zero correlation*—legs would get longer while teeth changed not at all. Other correlations can be negative—one measure increases while the other decreases. We begin to lose neurons at a distressingly early age, and they are not replaced. Thus, the relationship between leg length and number of neurons after mid-childhood represents *negative correlation*—leg length increases while number of neurons decreases. Notice that I have said nothing about causality. We do not know why these correlations exist or do not exist, only that they are present or not present.

The standard measure of correlation is called Pearson's product moment correlation coefficient or, for short, simply the correlation coefficient, symbolized as *r*. The correlation coefficient ranges from +1 for perfect positive correlation, to 0 for no correlation, to −1 for perfect negative correlation.*

In rough terms, *r* measures the shape of an ellipse of plotted points (see Fig. 6.1). Very skinny ellipses represent high correlations—the skinniest of all, a straight line, reflects an *r* of 1.0. Fat ellipses represent lower correlations, and the fattest of all, a circle, reflects zero correlation (increase in one measure permits no prediction about whether the other will increase, decrease, or remain the same).

The correlation coefficient, though easily calculated, has been plagued by errors of interpretation. These can be illustrated by example. Suppose that I plot arm length vs. leg length during the growth of a child. I will obtain a high correlation with two interesting implications. First, I have achieved *simplification*. I began with two dimensions (leg and arm length), which I have now, effectively, reduced to one. Since the correlation is so strong, we may say that the line itself (a single dimension) represents nearly all the information originally supplied as two dimensions. Secondly, I can, in this case, make a reasonable inference about the *cause* of this reduc-

*Pearson's *r* is not an appropriate measure for all kinds of correlation, for it assesses only what statisticians call the intensity of linear relationship between two measures—the tendency for all points to fall on a single straight line. Other relationships of strict dependence will not achieve a value of 1.0 for *r*. If, for example, each increase of 2 units in one variable were matched by an increase in 2^2 units in the other variable, *r* would be less than 1.0, even though the two variables might be perfectly "correlated" in the vernacular sense. Their plot would be a parabola, not a straight line, and Pearson's *r* measures the intensity of linear resemblance.

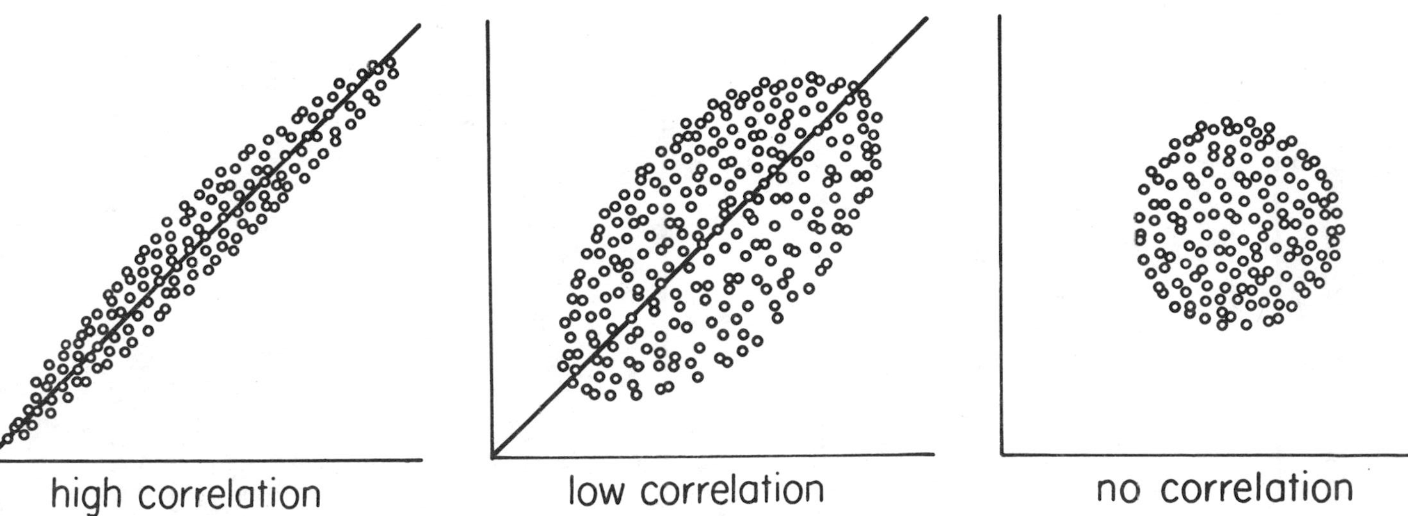

6 • 1 Strength of correlation as a function of the shape of an ellipse of points. The more elongate the ellipse, the higher the correlation.

tion to one dimension. Arm and leg length are tightly correlated because they are both partial measures of an underlying biological phenomenon, namely growth itself.

Yet, lest anyone become too hopeful that correlation represents a magic method for the unambiguous identification of cause, consider the relationship between my age and the price of gasoline during the past ten years. The correlation is nearly perfect, but no one would suggest any assignment of cause. The fact of correlation implies nothing about cause. It is not even true that intense correlations are more likely to represent cause than weak ones, for the correlation of my age with the price of gasoline is nearly 1.0. I spoke of cause for arm and leg lengths not because their correlation was high, but because I know something about the biology of the situation. The inference of cause must come from somewhere else, not from the simple fact of correlation—though an unexpected correlation may lead us to search for causes so long as we remember that we may not find them. The vast majority of correlations in our world are, without doubt, noncasual. Anything that has been decreasing steadily during the past few years will be strongly correlated with the distance between the earth and Halley's comet (which has also been decreasing of late)—but even the most dedicated astrologer would not discern causality in most of these relationships. The invalid assumption that correlation implies cause is probably among the two or three most serious and common errors of human reasoning.

Few people would be fooled by such a reductio ad absurdum as the age-gas correlation. But consider an intermediate case. I am given a table of data showing how far twenty children can hit and throw a baseball. I graph these data and calculate a high *r*. Most people, I think, would share my intuition that this is not a meaningless correlation; yet in the absence of further information, the correlation itself teaches me nothing about underlying causes. For I can suggest at least three different and reasonable causal interpretations for the correlation (and the true reason is probably some combination of them):

1. The children are simply of different ages, and older children can hit and throw farther.

2. The differences represent variation in practice and training. Some children are Little League stars and can tell you the year that

Rogers Hornsby hit .424 (1924—I was a bratty little kid like that); others know Billy Martin only as a figure in Lite beer commercials.

3. The differences represent disparities in native ability that cannot be erased even by intense training. (The situation would be even more complex if the sample included both boys and girls of conventional upbringing. The correlation might then be attributed primarily to a fourth cause—sexual differences; and we would have to worry, in addition, about the cause of the sexual difference: training, inborn constitution, or some combination of nature and nurture).

In summary, most correlations are noncausal; when correlations are causal, the fact and strength of the correlation rarely specifies the nature of the cause.

Correlation in more than two dimensions

These two-dimensional examples are easy to grasp (however difficult they are to interpret). But what of correlations among more than two measures? A body is composed of many parts, not just arms and legs, and we may want to know how several measures interact during growth. Suppose, for simplicity, that we add just one more measure, head length, to make a three-dimensional system. We may now depict the correlation structure among the three measures in two ways:

1. We may gather all correlation coefficients between pairs of measures into a single table, or *matrix* of correlation coefficients (Fig. 6.2). The line from upper left to lower right records the necessarily perfect correlation of each variable with itself. It is called the principal diagonal, and all correlations along it are 1.0. The matrix is symmetrical around the principal diagonal, since the correlation of measure 1 with measure 2 is the same as the correlation of 2 with 1. Thus, the three values either above or below the principal diagonal are the correlations we seek: arm with leg, arm with head, and leg with head.

2. We may plot the points for all individuals onto a three-dimensional graph (Fig. 6.3). Since the correlations are all positive, the points are oriented as an ellipsoid (or football). (In two dimensions, they formed an ellipse.) A line running along the major axis of the football expresses the strong positive correlations between all measures.

	arm	leg	head
arm	1.0	0.91	0.72
leg	0.91	1.0	0.63
head	0.72	0.63	1.0

6 • 2 A correlation matrix for three measurements.

6 • 3 A three-dimensional graph showing the correlations for three measurements.

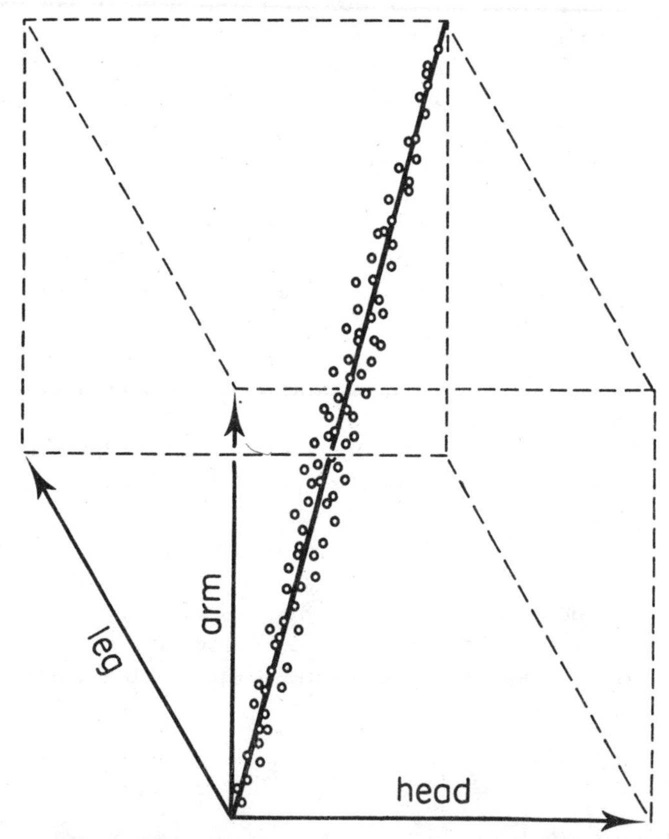

We can grasp the three-dimensional case, both mentally and pictorially. But what about 20 dimensions, or 100? If we measured 100 parts of a growing body, our correlation matrix would contain 10,000 items. To plot this information, we would have to work in a 100-dimensional space, with 100 mutually perpendicular axes representing the original measures. Although these 100 axes present no mathematical problem (they form, in technical terms, a hyperspace), we cannot plot them in our three-dimensional Euclidian world.

These 100 measures of a growing body probably do not represent 100 different biological phenomena. Just as most of the information in our three-dimensional example could be resolved into a single dimension (the long axis of the football), so might our 100 measures be simplified into fewer dimensions. We will lose some information in the process to be sure—as we did when we collapsed the long and skinny football, still a three-dimensional structure, into the single line representing its long axis. But we may be willing to accept this loss in exchange for simplification and for the possibility of interpreting the dimensions that we do retain in biological terms.

Factor analysis and its goals

With this example, we come to the heart of what *factor analysis* attempts to do. Factor analysis is a mathematical technique for reducing a complex system of correlations into fewer dimensions. It works, literally, by factoring a matrix, usually a matrix of correlation coefficients. (Remember the high-school algebra exercise called "factoring," where you simplified horrendous expressions by removing common multipliers of all terms?) Geometrically, the process of factoring amounts to placing axes through a football of points. In the 100-dimensional case, we are not likely to recover enough information on a single line down the hyperfootball's long axis—a line called the *first principal component.* We will need additional axes. By convention, we represent the second dimension by a line *perpendicular* to the first principal component. This second axis, or *second principal component,* is defined as the line that resolves more of the remaining variation than any other line that could be drawn perpendicular to the first principal component. If, for example, the hyperfootball were squashed flat like a flounder, the first principal component would run through the middle, from head to tail, and the second also through the middle, but from side to side. Subsequent lines would be perpendicular to all previous axes, and would resolve a steadily decreasing amount of remaining variation. We might find that five principal components resolve almost all the variation in our hyperfootball—that is, the hyperfootball drawn in 5 dimensions looks sufficiently like the original to satisfy us, just as a pizza or a flounder drawn in two dimensions may express all the information we need, even though both original objects contain three dimensions. If we elect to stop at 5 dimensions, we may achieve a considerable simplification at the acceptable price of minimal loss of information. We can grasp the 5 dimensions conceptually; we may even be able to interpret them biologically.

Since factoring is performed on a correlation matrix, I shall use a geometrical representation of the correlation coefficients themselves in order to explain better how the technique operates. The original measures may be represented as vectors of unit length,*

*(Footnote for aficionados—others may safely skip.) Here, I am technically discussing a procedure called "principal components analysis," not quite the same thing as factor analysis. In principal components analysis, we preserve all information in the original measures and fit new axes to them by the same criterion used in factor analysis in principal components orientation—that is, the first axis explains more data than any other axis could and subsequent axes lie at right angles to all other axes and encompass steadily decreasing amounts of information. In true factor analysis, we decide beforehand (by various procedures) not to include all information on our factor axes. But the two techniques—true factor analysis in principal components orientation and principal components analysis—play the same conceptual role and differ only in mode of calculation. In both, the first axis (Spearman's *g* for intelligence tests) is a "best fit" dimension that resolves more information in a set of vectors than any other axis could.

During the past decade or so, semantic confusion has spread in statistical circles through a tendency to restrict the term "factor analysis" only to the rotations of axes usually performed after the calculation of principal components, and to extend the term "principal components analysis" both to true principal components analysis (all information retained) and to factor analysis done in principal components orientation (reduced dimensionality and loss of information). This shift in definition is completely out of keeping with the history of the subject and terms. Spearman, Burt, and hosts of other psychometricians worked for decades in this area before Thurstone and others invented axial rotations. They performed all their calculations in the principal components orientation, and they called themselves "factor analysts." I continue, therefore, to use the term "factor analysis" in its original sense to include any orientation of axes—principal components or rotated, orthogonal or oblique.

I will also use a common, if somewhat sloppy, shorthand in discussing what radiating from a common point. If two measures are highly correlated, their vectors lie close to each other. The cosine of the angle between any two vectors records the correlation coefficient between them. If two vectors overlap, their correlation is perfect, or 1.0; the cosine of 0° is 1.0. If two vectors lie at right angles, they are completely independent, with a correlation of zero; the cosine of 90° is zero. If two vectors point in opposite directions, their correlation is perfectly negative, or −1.0; the cosine of 180° is −1.0. A matrix of high positive correlation coefficients will be represented by a cluster of vectors, each separated from each other vector by a small acute angle (Fig. 6.4). When we factor such a cluster into fewer dimensions by computing principal components, we choose as our first component the axis of maximal resolving power, a kind of grand average among all vectors. We assess resolving power by projecting each vector onto the axis. This is done by drawing a line from the tip of the vector to the axis, perpendicular to the axis. The ratio of projected length on the axis to the actual length of the vector itself measures the percentage of a vector's information resolved by the axis. (This is difficult to express verbally, but I think that Figure 6.5 will dispel confusion.) If a vector lies near the axis, it is highly resolved and the axis encompasses most of its information. As a vector moves away from the axis toward a maximal separation of 90°, the axis resolves less and less of it.

We position the first principal component (or axis) so that it resolves more information among all the vectors than any other axis could. For our matrix of high positive correlation coefficients, represented by a set of tightly clustered vectors, the first principal component runs through the middle of the set (Fig. 6.4). The second principal component lies at right angles to the first and resolves a maximal amount of remaining information. But if the first component has already resolved most of the information in all the vectors, then the second and subsequent principal axes can only deal with the small amount of information that remains (Fig. 6.4).

---

factor axes do. Technically, factor axes resolve variance in original measures. I will, as is often done, speak of them as "explaining" or "resolving" information—as they do in the vernacular (though not in the technical) sense of information. That is, when the vector of an original variable projects strongly on a set of factor axes, little of its variance lies unresolved in higher dimensions outside the system of factor axes.

Such systems of high positive correlation are found frequently in nature. In my own first study in factor analysis, for example, I considered fourteen measurements on the bones of twenty-two species of pelycosaurian reptiles (the fossil beasts with the sails on their backs, often confused with dinosaurs, but actually the ancestors of mammals). My first principal component resolved 97.1 per-

6•4 Geometric representation of correlations among eight tests when all correlation coefficients are high and positive. The first principal component, labeled 1, lies close to all the vectors, while the second principal component, labeled 2, lies at right angles to the first and does not explain much information in the vectors.

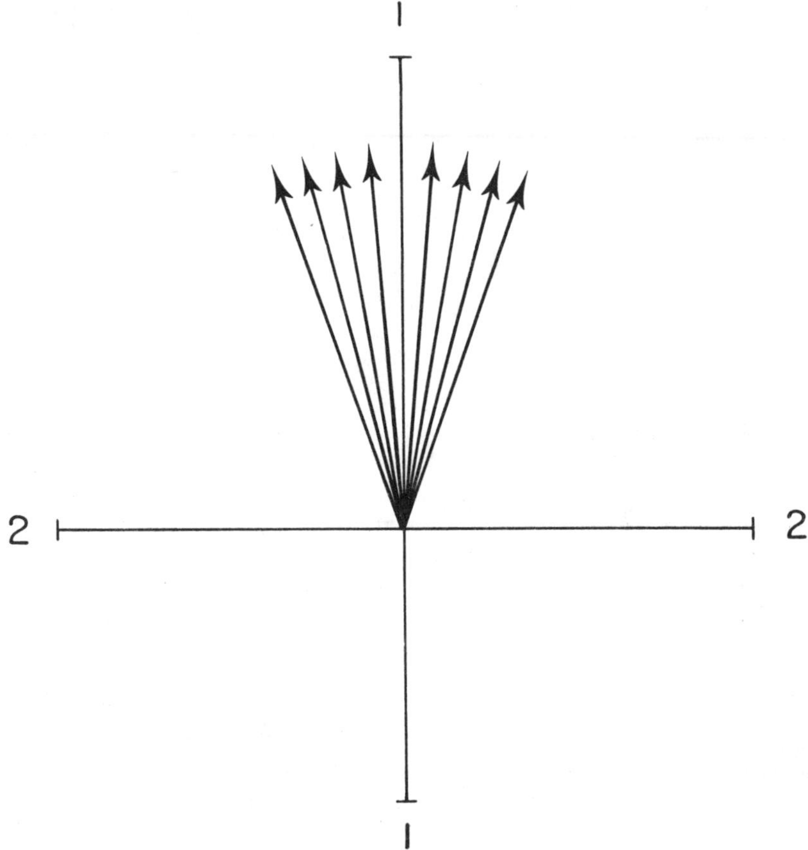

cent of the information in all fourteen vectors, leaving only 2.9 percent for subsequent axes. My fourteen vectors formed an extremely tight swarm (all practically overlapping); the first axis went through the middle of the swarm. My pelycosaurs ranged in body length from less than two to more than eleven feet. They all look pretty much alike, and big animals have larger measures for all fourteen bones. All correlation coefficients of bones with other bones are very high; in fact, the lowest is still a whopping 0.912.

6•5 Computing the amount of information in a vector explained by an axis. Draw a line from the tip of the vector to the axis, perpendicular to the axis. The amount of information resolved by the axis is the ratio of the projected length on the axis to the true length of the vector. If a vector lies close to the axis, then this ratio is high and most of the information in the vector is resolved by the axis. Vector AB lies close to the axis and the ratio of the projection AB′ to the vector itself, AB, is high. Vector AC lies far from the axis and the ratio of its projected length AC′ to the vector itself, AC, is low.

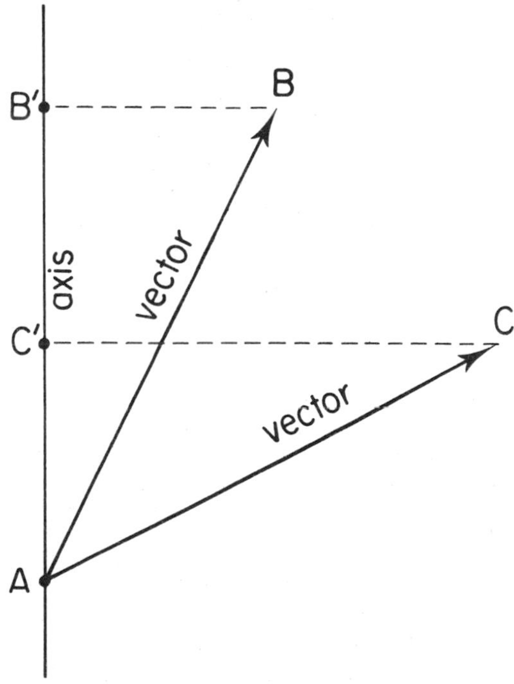

Scarcely surprising. After all, large animals have large bones, and small animals small bones. I can interpret my first principal component as an abstracted size factor, thus reducing (with minimal loss of information) my fourteen original measurements into a single dimension interpreted as increasing body size. In this case, factor analysis has achieved both *simplification* by reduction of dimensions (from fourteen to effectively one), and *explanation* by reasonable biological interpretation of the first axis as a size factor.

But—and here comes an enormous but—before we rejoice and extol factor analysis as a panacea for understanding complex systems of correlation, we should recognize that it is subject to the same cautions and objections previously examined for the correlation coefficients themselves. I consider two major problems in the following sections.

The error of reification

The first principal component is a mathematical abstraction that can be calculated for any matrix of correlation coefficients; it is not a "thing" with physical reality. Factorists have often fallen prey to a temptation for *reification*—for awarding *physical meaning* to all strong principal components. Sometimes this is justified; I believe that I can make a good case for interpreting my first pelycosaurian axis as a size factor. But such a claim can never arise from the mathematics alone, only from additional knowledge of the physical nature of the measures themselves. For nonsensical systems of correlation have principal components as well, and they may resolve more information than meaningful components do in other systems. A factor analysis for a five-by-five correlation matrix of my age, the population of Mexico, the price of swiss cheese, my pet turtle's weight, and the average distance between galaxies during the past ten years will yield a strong first principal component. This component—since all the correlations are so strongly positive—will probably resolve as high a percentage of information as the first axis in my study of pelycosaurs. It will also have no enlightening physical meaning whatever.

In studies of intelligence, factor analysis has been applied to matrices of correlation among mental tests. Ten tests may, for example, be given to each of one hundred people. Each meaningful entry in the ten-by-ten correlation matrix is a correlation coefficient between scores on two tests taken by each of the one hundred persons. We have known since the early days of mental testing—and it should surprise no one—that most of these correlation coefficients are positive: that is, people who score highly on one kind of test tend, on average, to score highly on others as well. Most correlation matrices for mental tests contain a preponderance of positive entries. This basic observation served as the starting point for factor analysis. Charles Spearman virtually invented the technique in 1904 as a device for inferring causes from correlation matrices of mental tests.

Since most correlation coefficients in the matrix are positive, factor analysis must yield a reasonably strong first principal component. Spearman calculated such a component indirectly in 1904 and then made the cardinal invalid inference that has plagued factor analysis ever since. He reified it as an "entity" and tried to give it an unambiguous causal interpretation. He called it *g*, or general intelligence, and imagined that he had identified a unitary quality underlying all cognitive mental activity—a quality that could be expressed as a single number and used to rank people on a unilinear scale of intellectual worth.

Spearman's *g*—the first principal component of the correlation matrix of mental tests—never attains the predominant role that a first component plays in many growth studies (as in my pelycosaurs). At best, *g* resolves 50 to 60 percent of all information in the matrix of tests. Correlations between tests are usually far weaker than correlations between two parts of a growing body. In most cases, the highest correlation in a matrix of tests does not come close to reaching the *lowest* value in my pelycosaur matrix—0.912.

Although *g* never matches the strength of a first principal component of some growth studies, I do not regard its fair resolving power as accidental. Causal reasons lie behind the positive correlations of most mental tests. But what reasons? We cannot infer the reasons from a strong first principal component any more than we can induce the cause of a single correlation coefficient from its magnitude. We cannot reify *g* as a "thing" unless we have convincing, independent information beyond the fact of correlation itself.

The situation for mental tests resembles the hypothetical case I presented earlier of correlation between throwing and hitting a baseball. The relationship is strong and we have a right to regard it as nonaccidental. But we cannot infer the cause from the correlation, and the cause is certainly complex.

Spearman's *g* is particularly subject to ambiguity in interpretation, if only because the two most contradictory causal hypotheses are both fully consistent with it: 1) that it reflects an inherited level of mental acuity (some people do well on most tests because they are born smarter); or 2) that it records environmental advantages and deficits (some people do well on most tests because they are well schooled, grew up with enough to eat, books in the home, and loving parents). If the simple existence of *g* can be theoretically interpreted in either a purely hereditarian or purely environmentalist way, then its mere presence—even its reasonable strength—cannot justly lead to any reification at all. The temptation to reify is powerful. The idea that we have detected something "underlying" the externalities of a large set of correlation coefficients, something perhaps more real than the superficial measurements themselves, can be intoxicating. It is Plato's essence, the abstract, eternal reality underlying superficial appearances. But it is a temptation that we must resist, for it reflects an ancient prejudice of thought, not a truth of nature.

Rotation and the nonnecessity of principal components

Another, more technical, argument clearly demonstrates why principal components cannot be automatically reified as causal entities. If principal components represented the only way to simplify a correlation matrix, then some special status for them might be legitimately sought. But they represent only one method among many for inserting axes into a multidimensional space. Principal components have a definite geometric arrangement, specified by the criterion used to construct them—that the first principal component shall resolve a maximal amount of information in a set of vectors and that subsequent components shall all be mutually perpendicular. But there is nothing sacrosánct about this criterion; vectors may be resolved into any set of axes placed within their space. Principal components provide insight in some cases, but other criteria are often more useful.

Consider the following situation, in which another scheme for placing axes might be preferred. In Figure 6.6 I show correlations between four mental tests, two of verbal and two of arithmetical aptitude. Two "clusters" are evident, even though all tests are positively correlated. Suppose that we wish to identify these clusters by factor analysis. If we use principal components, we may not recognize them at all. The first principal component (Spearman's *g*) goes right up the middle, between the two clusters. It lies close to no vector and resolves an approximately equal amount of each, thereby masking the existence of verbal and arithmetic clusters. Is this component an entity? Does a "general intelligence" exist? Or is *g*, in this case, merely a meaningless average based on the invalid amalgamation of two types of information?

We may pick up verbal and arithmetic clusters on the second principal component (called a "bipolar factor" because some projections upon it will be positive and others negative when vectors lie on both sides of the first principal component). In this case, verbal tests project on the negative side of the second component, and arithmetic tests on the positive side. But we may fail to detect these clusters altogether if the first principal component dominates all vectors. For projections on the second component will then be small, and the pattern can easily be lost (see Fig. 6.6).

During the 1930s factorists developed methods to treat this dilemma and to recognize clusters of vectors that principal components often obscured. They did this by rotating factor axes from the principal components orientation to new positions. The rotations, established by several criteria, had as their common aim the positioning of axes near clusters. In Figure 6.7, for example, we use the criterion: place axes near vectors occupying extreme or outlying positions in the total set. If we now resolve all vectors into these rotated axes, we detect the clusters easily; for arithmetic tests project high on rotated axis 1 and low on rotated axis 2, while verbal tests project high on 2 and low on 1. Moreover, *g has disappeared.* We no longer find a "general factor" of intelligence, nothing that can be reified as a single number expresssing overall ability. Yet we have lost no information. The two rotated axes resolve as much information in the four vectors as did the two principal components. They simply distribute the same information differently upon the resolving axes. How can we argue that *g* has any claim to reified status as an entity if it represents but one of numerous possible ways to position axes within a set of vectors?

In short, factor analysis simplifies large sets of data by reducing dimensionality and trading some loss of information for the recognition of ordered structure in fewer dimensions. As a tool for simplification, it has proved its great value in many disciplines. But many factorists have gone beyond simplification, and tried to define factors as causal entities. This error of reification has plagued the technique since its inception. It was "present at the creation" since Spearman invented factor analysis to study the correlation matrix of mental tests and then reified his principal component as *g* or innate, general intelligence. Factor analysis may help us to understand causes by directing us to information beyond the

6 • 6 A principal components analysis of four mental tests. All correlations are high and the first principal component, Spearman's *g*, expresses the overall correlation. But the group factors for verbal and mathematical aptitude are not well resolved in this style of analysis.

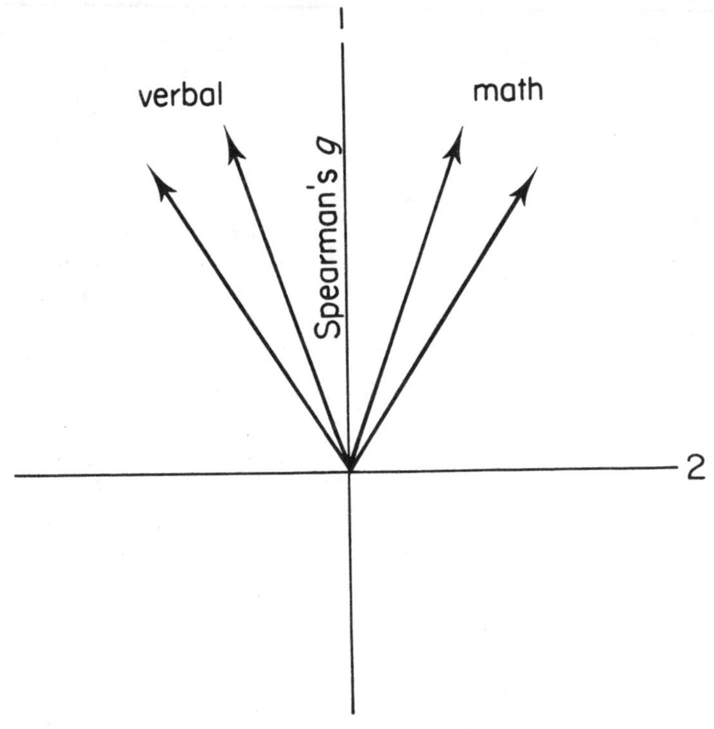

mathematics of correlation. But factors, by themselves, are neither things nor causes; they are mathematical abstractions. Since the same set of vectors (see Figs. 6.6, 6.7) can be partitioned into *g* and a small residual axis, or into two axes of equal strength that identify verbal and arithmetical clusters and dispense with *g* entirely, we cannot claim that Spearman's "general intelligence" is an ineluctable entity necessarily underlying and causing the correlations among mental tests. Even if we choose to defend *g* as a nonaccidental result, neither its strength nor its geometric position can specify what it means in causal terms—if only because its features are equally consistent with extreme hereditarian and extreme environmentalist views of intelligence.

6•7 Rotated factor axes for the same four mental tests depicted in Fig. 6.6. Axes are now placed near vectors lying at the periphery of the cluster. The group factors for verbal and mathematical aptitude are now well identified (see high projections on the axes indicated by dots), but *g* has disappeared.

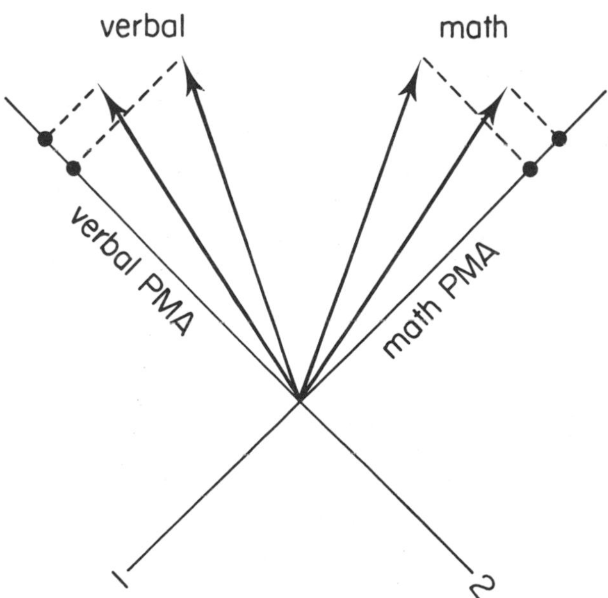

Charles Spearman and general intelligence

The two-factor theory

Correlation coefficients are now about as ubiquitous and unsurprising as cockroaches in New York City. Even the cheapest pocket calculators produce correlation coefficients with the press of a button or the pass of a magnetic tape. However indispensable, they are taken for granted as automatic accouterments of any statistical analysis that deals with more than one measure. In such a context, we easily forget that they were once hailed as a breakthrough in research, as a new and exciting tool for discovering underlying structure in tables of raw measures. We can sense this excitement in reading early papers of the great American biologist and statistician Raymond Pearl (see Pearl, 1905 and 1906, and Pearl and Fuller, 1905). Pearl completed his doctorate at the turn of the century and then proceeded, like a happy boy with a gleaming new toy, to correlate everything in sight, from the lengths of earth worms vs. the number of their body segments (where he found no correlation and assumed that increasing length reflects larger, rather than more, segments), to size of the human head vs. intelligence (where he found a very small correlation, but attributed it to the indirect effect of better nutrition).

Charles Spearman, an eminent psychologist and fine statistician as well* began to study correlations between mental tests during these heady times. If two mental tests are given to a large number of people, Spearman noted, the correlation coefficient between them is nearly always positive. Spearman pondered this result and wondered what higher generality it implied. The positive correlations clearly indicated that each test did not measure an independent attribute of mental functioning. Some simpler structure lay behind the pervasive positive correlations; but what structure? Spearman imagined two alternatives. First, the positive correlations might reduce to a small set of independent attributes—the "faculties" of the phrenologists and other schools of early psychology. Perhaps the mind had separate "compartments" for arithmetic, verbal, and spatial aptitudes, for example. Spearman called such

*Spearman took a special interest in problems of correlation and invented a measure that probably ranks second in use to Pearson's r as a measure of association between two variables—the so-called Spearman's rank-correlation coefficient.

theories of intelligence "oligarchic." Second, the positive correlations might reduce to a single, underlying general factor—a notion that Spearman called "monarchic." In either case, Spearman recognized that the underlying factors—be they few (oligarchic) or single (monarchic)—would not encompass all information in a matrix of positive correlation coefficients for a large number of mental tests. A "residual variance" would remain—information peculiar to each test and not related to any other. In other words, each test would have its "anarchic" component. Spearman called the residual variance of each test its *s*, or specific information. Thus, Spearman reasoned, a study of underlying structure might lead to a "two-factor theory" in which each test contained some specific information (its *s*) and also reflected the operation of a single, underlying factor, which Spearman called *g*, or general intelligence. Or each test might include its specific information and also record one or several among a set of independent, underlying faculties—a many-factor theory. If the simplest two-factor theory held, then all common attributes of intelligence would reduce to a single underlying entity—a true "general intelligence" that might be measured for each person and might afford an unambiguous criterion for ranking in terms of mental worth.

Charles Spearman developed factor analysis—still the most important technique in modern multivariate statistics—as a procedure for deciding between the two- vs. the many-factor theory by determining whether the common variance in a matrix of correlation coefficients could be reduced to a single "general" factor, or only to several independent "group" factors. He found but a single "intelligence," opted for the two-factor theory, and, in 1904, published a paper that later won this assessment from a man who opposed its major result: "No single event in the history of mental testing has proved to be of such momentous importance as Spearman's proposal of his famous two-factor theory" (Guilford, 1936, p. 155). Elated, and with characteristic immodesty, Spearman gave his 1904 paper a heroic title: "General Intelligence Objectively Measured and Determined." Ten years later (1914, p. 237), he exulted: "The future of research into the inheritance of ability must center on the theory of 'two factors.' This alone seems capable of reducing the bewildering chaos of facts to a perspicuous orderliness. By its means, the problems are rendered clear; in many respects, their answers are already foreshadowed; and everywhere, they are rendered susceptible of eventual decisive solution."

The method of tetrad differences

In his original work, Spearman did not use the method of principal components described on pp. 245–248. Instead, he developed a simpler, though tedious, procedure better suited for a precomputer age when all calculations had to be performed by hand.* He computed the entire matrix of correlation coefficients between all pairs of tests, took all possible groupings of four measures and computed for each a number that he called the "tetrad difference." Consider the following example as an attempt to define the tetrad difference and to explain how Spearman used it to test whether the common variance of his matrix could be reduced to a single general factor, or only to several group factors.

Suppose that we wish to compute the tetrad difference for four measures taken on a series of mice ranging in age from babies to adults—leg length, leg width, tail length, and tail width. We compute all correlation coefficients between pairs of variables and find, unsurprisingly, that all are positive—as mice grow, their parts get larger. But we would like to know whether the common variance in the positive correlations all reflects a single general factor—growth itself—or whether two separate components of growth must be identified—in this case, a leg factor and a tail factor, or a length factor and a width factor. Spearman gives the following formula for the tetrad difference

$$r_{13} \times r_{24} - r_{23} \times r_{14}$$

where r is the correlation coefficient and the two subscripts represent the two measures being correlated (in this case, 1 is leg length, 2 is leg width, 3 is tail length and 4 is tail width—so that r_{13} is the correlation coefficient between the first and the third measure, or between leg length and tail length). In our example, the tetrad difference is

(leg length and tail length) × (leg width and tail width) −

(leg width and tail length) × (leg length and tail width)

*The *g* calculated by the tetrad formula is conceptually equivalent and mathematically almost equivalent to the first principal component described on pp. 245–248 and used in modern factor analysis.

Spearman argued that tetrad differences of zero imply the existence of a single general factor while either positive or negative values indicate that group factors must be recognized. Suppose, for example, that group factors for general body length and general body width govern the growth of mice. In this case, we would get a high positive value for the tetrad difference because the correlation coefficients of a length with another length or a width with another width would tend to be higher than correlation coefficients of a width with a length. (Note that the left-hand side of the tetrad equation includes only lengths with lengths or widths with widths, while the right-hand side includes only lengths with widths.) But if only a single, general growth factor regulates the size of mice, then lengths with widths should show as high a correlation as lengths with lengths or widths with widths—and the tetrad difference should be zero. Fig. 6.8 shows a hypothetical correlation matrix for the four measures that yields a tetrad difference of zero (values taken from Spearman's example in another context, 1927, p. 74). Fig. 6.8 also shows a different hypothetical matrix yielding a positive tetrad difference and a conclusion (if other tetrads show the same pattern) that group factors for length and width must be recognized.

The top matrix of Fig. 6.8 illustrates another important point that reverberates throughout the history of factor analysis in psychology. Note that, although the tetrad difference is zero, the correlation coefficients need not be (and almost invariably are not) equal. In this case, leg width with leg length gives a correlation of 0.80, while tail width with tail length yields only 0.18. These differences reflect varying "saturations" with *g*, the single general factor when the tetrad differences are zero. Leg measures have higher saturations than tail measures—that is, they are closer to *g*, or reflect it better (in modern terms, they lie closer to the first principal component in geometric representations like Fig. 6.6). Tail measures do not load strongly on *g*.* They contain little common variance and must be explained primarily by their *s*—the information unique to each measure. Moving now to mental tests: if *g* represents general intelligence, then mental tests most saturated with

*The terms "saturation" and "loading" refer to the correlation between a test and a factor axis. If a test "loads" strongly on a factor then most of its information is explained by the factor.

	LL	LW	TL	TW
LL	1.0			
LW	0.80	1.0		
TL	0.60	0.48	1.0	
TW	0.30	0.24	0.18	1.0

Tetrad difference:

0.60 x 0.24 - 0.48 x 0.3

0.144 - 0.144 = 0

no group factors

	LL	LW	TL	TW
LL	1.0			
LW	0.80	1.0		
TL	0.40	0.20	1.0	
TW	0.20	0.40	0.50	1.0

Tetrad difference:

0.40 x 0.40 - 0.20 x 0.20

0.16 - 0.04 = 0.12

group factors for lengths and widths

6 • 8 Tetrad differences of zero (above) and a positive value (below) from hypothetical correlation matrices for four measurements: LL = leg length, LW = leg width, TL = tail length, and TW = tail width. The positive tetrad difference indicates the existence of group factors for lengths and widths.

g are the best surrogates for general intelligence, while tests with low *g*-loadings (and high *s* values) cannot serve as good measures of general mental worth. Strength of *g*-loading becomes the criterion for determining whether or not a particular mental test (IQ, for example) is a good measure of general intelligence.

Spearman's tetrad procedure is very laborious when the correlation matrix includes a large number of tests. Each tetrad difference must be calculated separately. If the common variance reflects but a single general factor, then the tetrads should equal zero. But, as in any statistical procedure, not all cases meet the expected value (half heads and half tails is the expectation in coin flipping, but you will flip six heads in a row about once in sixty-four series of six flips). Some calculated tetrad differences will be positive or negative even when a single *g* exists and the expected value is zero. Thus, Spearman computed all tetrad differences and looked for normal frequency distributions with a mean tetrad difference of zero as his test for the existence of *g*.

Spearman's g *and the great instauration of psychology*

Charles Spearman computed all his tetrads, found a distribution close enough to normal with a mean close enough to zero, and proclaimed that the common variance in mental tests recorded but a single underlying factor—Spearman's *g*, or general intelligence. Spearman did not hide his pleasure, for he felt that he had discovered the elusive entity that would make psychology a true science. He had found the innate essence of intelligence, the reality underlying all the superficial and inadequate measures devised to search for it. Spearman's *g* would be the philosopher's stone of psychology, its hard, quantifiable "thing"—a fundamental particle that would pave the way for an exact science as firm and as basic as physics.

In his 1904 paper, Spearman proclaimed the ubiquity of *g* in all processes deemed intellectual: "All branches of intellectual activity have in common one fundamental function . . . whereas the remaining or specific elements seem in every case to be wholly different from that in all the others. . . . This *g*, far from being confined to some small set of abilities whose intercorrelations have actually been measured and drawn up in some particular table, may enter into all abilities whatsoever."

The conventional school subjects, insofar as they reflect aptitude rather than the simple acquisition of information, merely peer through a dark glass at the single essence inside: "All examination in the different sensory, school, and other specific faculties may be considered as so many independently obtained estimates of the one great common Intellective Function" (1904, p. 273). Thus Spearman tried to resolve a traditional dilemma of conventional education for the British elite: why should training in the classics make a better soldier or a statesman? "Instead of continuing ineffectively to protest that high marks in Greek syntax are no test as to the capacity of men to command troops or to administer provinces, we shall at last actually determine the precise accuracy of the various means of measuring General Intelligence" (1904, p. 277). In place of fruitless argument, one has simply to determine the *g*-loading of Latin grammar and military acuity. If both lie close to *g*, then skill in conjugation may be a good estimate of future ability to command.

There are different styles of doing science, all legitimate and partially valid. The beetle taxonomist who delights in noting the peculiarities of each new species may have little interest in reduction, synthesis, or in probing for the essence of "beetleness"—if such exists! At an opposite extreme, occupied by Spearman, the externalities of this world are only superficial guides to a simpler, underlying reality. In a popular image (though some professionals would abjure it), physics is the ultimate science of reduction to basic and quantifiable causes that generate the apparent complexity of our material world. Reductionists like Spearman, who work in the so-called soft sciences of organismic biology, psychology, or sociology, have often suffered from "physics envy." They have strived to practice their science according to their clouded vision of physics—to search for simplifying laws and basic particles. Spearman described his deepest hopes for a science of cognition (1923, p. 30):

> Deeper than the uniformities of occurrence which are noticeable even without its aid, it [science] discovers others more abstruse, but correspondingly more comprehensive, upon which the name of laws is bestowed. . . . When we look around for any approach to this ideal, something of the sort can actually be found in the science of physics as based on the three primary laws of motion. Coordinate with this *physica corporis* [physics of bodies], then, we are today in search of a *physica animae* [physics of the soul].

With *g* as a quantified, fundamental particle, psychology could take its rightful place among the real sciences. "In these principles," he wrote in 1923 (p. 355), "we must venture to hope that the so long missing genuinely scientific foundation for psychology has at last been supplied, so that it can henceforward take its due place along with the other solidly founded sciences, even physics itself." Spearman called his work "a Copernican revolution in point of view" (1927, p. 411) and rejoiced that "this Cinderella among the sciences has made a bold bid for the level of triumphant physics itself" (1937, p. 21).

Spearman's g *and the theoretical justification of IQ*

Spearman, the theorist, the searcher for unity by reduction to underlying causes, often spoke in most unflattering terms about the stated intentions of IQ testers. He referred to IQ (1931) as "the mere average of sub-tests picked up and put together without rhyme or reason." He decried the dignification of this "gallimaufry of tests" with the name intelligence. In fact, though he had described his *g* as general intelligence in 1904, he later abandoned the word intelligence because endless arguments and inconsistent procedures of mental testers had plunged it into irremediable ambiguity (1927, p. 412; 1950, p. 67).

Yet it would be incorrect—indeed it would be precisely contrary to Spearman's view—to regard him as an opponent of IQ testing. He had contempt for the atheoretical empiricism of the testers, their tendency to construct tests by throwing apparently unrelated items together and then offering no justification for such a curious procedure beyond the claim that it yielded good results. Yet he did not deny that the Binet tests worked, and he rejoiced in the resuscitation of the subject thus produced: "By this one great investigation [the Binet scale] the whole scene was transformed. The recently despised tests were now introduced into every country with enthusiasm. And everywhere their practical application was brilliantly successful" (1914, p. 312).

What galled Spearman was his conviction that IQ testers were doing the right thing in amalgamating an array of disparate items into a single scale, but that they refused to recognize the theory behind such a procedure and continued to regard their work as rough-and-ready empiricism.

Spearman argued passionately that the justification for Binet testing lay with his own theory of a single *g* underlying all cognitive activity. IQ tests worked because, unbeknownst to their makers, they measured *g* with fair accuracy. Each individual test has a *g*-loading and its own specific information (or *s*), but *g*-loading varies from nearly zero to nearly 100 percent. Ironically, the most accurate measure of *g* will be the average score for a large collection of individual tests of the most diverse kind. Each measures *g* to some extent. The variety guarantees that *s*-factors of the individual tests will vary in all possible directions and cancel each other out. Only *g* will be left as the factor common to all tests. IQ works because it measures *g*.

> An explanation is at once supplied for the success of their extraordinary procedure of . . . pooling together tests of the most miscellaneous description. For if every performance depends on two factors, the one always varying randomly, while the other is constantly the same, it is clear that in the average the random variations will tend to neutralize one another, leaving the other, or constant factor, alone dominant (1914, p. 313; see also, 1923, p. 6, and 1927, p. 77).

Binet's "hotchpot of multitudinous measurements" was a correct theoretical decision, not only the intuitive guess of a skilled practitioner: "In such wise this principle of making a hotchpot, which might seem to be the most arbitrary and meaningless procedure imaginable, had really a profound theoretical basis and a supremely practical utility" (Spearman quoted in Tuddenham, 1962, p. 503).

Spearman's *g*, and its attendant claim that intelligence is a single, measurable entity, provided the only promising theoretical justification that hereditarian theories of IQ have ever had. As mental testing rose to prominence during the early twentieth century, it developed two traditions of research that Cyril Burt correctly identified in 1914 (p. 36) as correlational methods (factor analysis) and age-scale methods (IQ testing). Hearnshaw has recently made the same point in his biography of Burt (1979, p. 47): "The novelty of the 1900's was not in the concept of intelligence itself, but in its operational definition in terms of correlational techniques, and in the devising of practicable methods of measurement."

No one recognized better than Spearman the intimate connection between his model of factor analysis and hereditarian interpretations of IQ testing. In his 1914 *Eugenics Review* article, he prophesied the union of these two great traditions in mental testing: "Each of these two lines of investigation furnishes a peculiarly happy and indispensable support to the other. . . . Great as has been the value of the Simon-Binet tests, even when worked in theoretical darkness, their efficiency will be multiplied a thousand-fold when employed with a full light upon their essential nature and mechanism." When Spearman's style of factor analysis came under attack late in his career (see pp. 296–302), he defended *g* by citing it as the rationale for IQ: "Statistically, this determination is grounded on its extreme simpleness. Psychologically, it is credited with affording the sole base for such useful concepts as those of 'general ability,' or 'IQ' " (1939, p. 79).

To be sure, the professional testers did not always heed Spearman's plea for an adoption of *g* as the rationale for their work. Many testers abjured theory and continued to insist on practical utility as the justification for their efforts. But silence about theory does not connote an absence of theory. The reification of IQ as a biological entity has depended upon the conviction that Spearman's *g* measures a single, scalable, fundamental "thing" residing in the human brain. Many of the more theoretically inclined mental testers have taken this view (see Terman et al., 1917, p. 152). C. C. Brigham did not base his famous recantation solely upon a belated recognition that the army mental tests had considered patent measures of culture as inborn properties (pp. 232–233). He also pointed out that no strong, single *g* could be extracted from the combined tests, which, therefore, could not have been measures of intelligence after all (Brigham, 1930). And I will at least say this for Arthur Jensen: he recognizes that his hereditarian theory of IQ depends upon the validity of *g*, and devotes much of his recent book (1979) to a defense of Spearman's argument in its original form. A proper understanding of the conceptual errors in Spearman's formulation is a prerequisite for criticizing hereditarian claims about IQ at their fundamental level, not merely in the tangled minutiae of statistical procedures.

Spearman's reification of g

Spearman could not rest content with the idea that he had probed deeply under the empirical results of mental tests and found a single abstract factor underlying all performance. Nor could he achieve adequate satisfaction by identifying that factor with what we call intelligence itself.* Spearman felt compelled to ask more of his *g:* it must measure some physical property of the brain; it must be a "thing" in the most direct, material sense. Even if neurology had found no substance to identify with *g*, the brain's performance on mental tests proved that such a physical substrate must exist. Thus, caught up in physics envy again, Spearman described his own "adventurous step of deserting all actually observable phenomena of the mind and proceeding instead to invent an underlying something which—by analogy with physics—has been called mental energy" (1927, p. 89).

Spearman looked to the basic property of *g*—its influence in varying degree, upon mental operations—and tried to imagine what physical entity best fitted such behavior. What else, he argued, but a form of energy pervading the entire brain and activating a set of specific "engines," each with a definite locus. The more energy, the more general activation, the more intelligence. Spearman wrote (1923, p. 5):

> This continued tendency to success of the same person throughout all variations of both form and subject matter—that is to say, throughout all conscious aspects of cognition whatever—appears only explicable by some factor lying deeper than the phenomena of consciousness. And thus there emerges the concept of a hypothetical general and purely quantitative factor underlying all cognitive performances of any kind. . . . The factor was taken, pending further information, to consist in something of the nature of an "energy" or "power" which serves in common the whole cortex (or possibly, even, the whole nervous system)."

If *g* pervades the entire cortex as a general energy, then the *s*-factors for each test must have more definite locations. They must represent specific groups of neurons, activated in different ways by the energy identified with *g*. The *s*-factors, Spearman wrote (and not merely in metaphor), are engines fueled by a circulating *g*.

> Each different operation must necessarily be further served by some specific factor peculiar to it. For this factor also, a physiological substrate has been suggested, namely the particular group of neurons specially serving the particular kind of operation. These neural groups would thus

*At least in his early work. Later, as we have seen, he abandoned the word intelligence as a result of its maddening ambiguity in common usage. But he did not cease to regard *g* as the single cognitive essence that should be called intelligence, had not vernacular (and technical) confusion made such a mockery of the term.

> function as alternative "engines" into which the common supply of "energy" could be alternatively distributed. Successful action would always depend, partly on the potential of energy developed in the whole cortex, and partly on the efficiency of the specific group of neurons involved. The relative influence of these two factors could vary greatly according to the kind of operation; some kinds would depend more on the potential of the energy, others more on the efficiency of the engine (1923, pp. 5–6).

The differing *g*-loadings of tests had been provisionally explained: one mental operation might depend primarily upon the character of its engine (high *s* and low *g*-loading), another might owe its status to the amount of general energy involved in activating its enginc (high *g*-loading).

Spearman felt sure that he had discovered the basis of intelligence, so sure that he proclaimed his concept impervious to disproof. He expected that a physical energy corresponding with *g* would be found by physiologists: "There seem to be grounds for hoping that a material energy of the kind required by psychologists will some day actually be discovered" (1927, p. 407). In this discovery, Spearman proclaimed, "physiology will achieve the greatest of its triumphs" (1927, p. 408). But should no physical energy be found, still an energy there must be—but of a different sort:

> And should the worst arrive and the required physiological explanation remain to the end undiscoverable, the mental facts will none the less remain facts still. If they are such as to be best explained by the concept of an underlying energy, then this concept will have to undergo that which after all is only what has long been demanded by many of the best psychologists—it will have to be regarded as purely mental (1927, p. 408).

Spearman, in 1927 at least, never considered the obvious alternative: that his attempt to reify *g* might be invalid in the first place.

Throughout his career, Spearman tried to find other regularities of mental functioning that would validate his theory of general energy and specific engines. He enunciated (1927, p. 133) a "law of constant output" proclaiming that the cessation of any mental activity causes others of equal intensity to commence. Thus, he reasoned, general energy remains intact and must always be activating something. He found, on the other hand, that fatigue is "selectively transferred"—that is, tiring in one mental activity entails fatigue in some related areas, but not in others (1927, p. 318). Thus, fatigue cannot be attributed to "decrease in the supply of the general psycho-physiological energy," but must represent a build up of toxins that act selectively upon certain kinds of neurons. Fatigue, Spearman proclaimed, "primarily concerns not the energy but the engines" (1927, p. 318).

Yet, as we find so often in the history of mental testing, Spearman's doubts began to grow until he finally recanted in his last (posthumously published) book of 1950. He seemed to pass off the theory of energy and engines as a folly of youth (though he had defended it staunchly in middle age). He even abandoned the attempt to reify factors, recognizing belatedly that a mathematical abstraction need not correspond with a physical reality. The great theorist had entered the camp of his enemies and recast himself as a cautious empiricist (1950, p. 25):

> We are under no obligation to answer such questions as: whether "factors" have any "real" existence? do they admit of genuine "measurement"? does the notion of "ability" involve at bottom any kind of cause, or power? Or is it only intended for the purpose of bare description? . . . At their time and in their place such themes are doubtless well enough. The senior writer himself has indulged in them not a little. *Dulce est desipere in loco* [it is pleasant to act foolishly from time to time—a line from Horace]. But for the present purposes he has felt himself constrained to keep within the limits of barest empirical science. These he takes to be at bottom nothing but description and prediction. . . . The rest is mostly illumination by way of metaphor and similes.

The history of factor analysis is strewn with the wreckage of misguided attempts at reification. I do not deny that patterns of causality may have identifiable and underlying, physical reasons, and I do agree with Eysenck when he states (1953, p. 113): "Under certain circumstances, factors may be regarded as hypothetical causal influences underlying and determining the observed relationships between a set of variables. It is only when regarded in this light that they have interest and significance for psychology." My complaint lies with the practice of assuming that the mere existence of a factor, in itself, provides a license for causal speculation. Factorists have consistently warned against such an assumption, but our Platonic urges to discover underlying essences continue to prevail over proper caution. We can chuckle, with the beneficence of hindsight, at psychiatrist T. V. Moore who, in 1933, postulated definite genes for catatonic, deluded, manic, cognitive, and constitutional depression because his factor analysis grouped the supposed measures of these syndromes on separate axes (in Wolfle, 1940). Yet in 1972 two authors found an association of dairy production with florid vocalization on the tiny thirteenth axis of a nineteen-axis factor analysis for musical habits of various cultures—and then suggested "that this extra source of protein accounts for many cases of energetic vocalizing" (Lomax and Berkowitz, 1972, p. 232).

Automatic reification is invalid for two major reasons. First, as I discussed briefly on pp. 252–255 and will treat in full on pp. 296–317, no set of factors has any claim to exclusive concordance with the real world. Any matrix of positive correlation coefficients can be factored, as Spearman did, into *g* and a set of subsidiary factors or, as Thurstone did, into a set of "simple structure" factors that usually lack a single dominant direction. Since either solution resolves the same amount of information, they are equivalent in mathematical terms. Yet they lead to contrary psychological interpretations. How can we claim that one, or either, is a mirror of reality?

Second, any single set of factors can be interpreted in a variety of ways. Spearman read his strong *g* as evidence for a single reality underlying all cognitive mental activity, a general energy within the brain. Yet Spearman's most celebrated English colleague in factor analysis, Sir Godfrey Thomson, accepted Spearman's mathematical results but consistently chose to interpret them in an opposite manner. Spearman argued that the brain could be divided into a set of specific engines, fueled by a general energy. Thomson, using the same data, inferred that the brain has hardly any specialized structure at all. Nerve cells, he argued, either fire completely or not at all—they are either off or on, with no intermediary state. Every mental test samples a random array of neurons. Tests with high *g*-loadings catch many neurons in the active state; others, with low *g*-loadings, have simply sampled a smaller amount of unstructured brain. Thomson concluded (1939): "Far from being divided up into a few 'unitary factors,' the mind is a rich, comparatively undifferentiated complex of innumerable influences—on the physiological side an intricate network of possibilities of intercommunication." If the same mathematical pattern can yield such disparate interpretations, what claim can either have upon reality?

Spearman on the inheritance of g

Two of Spearman's primary claims appear in most hereditarian theories of mental testing: the identification of intelligence as a unitary "thing," and the inference of a physical substrate for it. But these claims do not complete the argument: a single, physical substance may achieve its variable strength through effects of environment and education, not from inborn differences. A more direct argument for the heritability of *g* must be made, and Spearman supplied it.

The identification of *g* and *s* with energy and engines again provided Spearman with his framework. He argued that the *s*-factors record training in education, but that the strength of a person's *g* reflects heredity alone. How can *g* be influenced by education, Spearman argued (1927, p. 392), if *g* ceases to increase by about age sixteen but education may continue indefinitely thereafter? How can *g* be altered by schooling if it measures what Spearman called *eduction* (or the ability to synthesize and draw connections) and not *retention* (the ability to learn facts and remember them)—when schools are in the business of imparting information? The engines can be stuffed full of information and shaped by training, but the brain's general energy is a consequence of its inborn structure:

> The effect of training is confined to the specific factor and does not touch the general one; physiologically speaking, certain neurons become habituated to particular kinds of action, but the free energy of the brain remains unaffected. . . . Though unquestionably the development of specific abilities is in large measure dependent upon environmental influences, that of general ability is almost wholly governed by heredity (1914, pp. 233–234).

IQ, as a measure of *g*, records an innate general intelligence; the marriage of the two great traditions in mental measurement (IQ testing and factor analysis) was consummated with the issue of heredity.

On the vexatious issue of group differences, Spearman's views accorded with the usual beliefs of leading western European male scientists at the time (see Fig. 6.9). Of blacks, he wrote (1927, p. 379), invoking *g* to interpret the army mental tests:

> On the average of all the tests, the colored were about two years behind the white; their inferiority extended through all ten tests, but it was most marked in just those which are known to be most saturated with *g*.

In other words, blacks performed most poorly on tests having strongest correlations with *g*, or innate general intelligence.

Of whites from southern and eastern Europe, Spearman wrote (1927, p. 379), praising the American Immigration Restriction Act of 1924:

> The general conclusion emphasized by nearly every investigator is that, as regards "intelligence," the Germanic stock has on the average a marked advantage over the South European. And this result would seem to have

6•9 Racist stereotype of a Jewish financier, reproduced from the first page of Spearman's 1914 article (see Bibliography). Spearman used this figure to criticize beliefs in group factors for such particular items of intellect, but its publication illustrates the acceptable attitudes of another age.

had vitally important practical consequences in shaping the recent very stringent American laws as to admission of immigrants.

Yet it would be incorrect to brand Spearman as an architect of the hereditarian theory for differences in intelligence among human groups. He supplied some important components, particularly the argument that intelligence is an innate, single, scorable "thing." He also held conventional views on the source of average differences in intelligence between races and national groups. But he did not stress the ineluctability of differences. In fact, he attributed sexual differences to training and social convention (1927, p. 229) and had rather little to say about social classes. Moreover, when discussing racial differences, he always coupled his hereditarian claim about average scores with an argument that the range of variation within any racial or national group greatly exceeds the small average difference between groups—so that many members of an "inferior" race will surpass the average intelligence of a "superior" group (1927, p. 380, for example).

Spearman also recognized the political force of hereditarian claims, though he did not abjure either the claim or the politics: "All great efforts to improve human beings by way of training are thwarted through the apathy of those who hold the sole feasible road to be that of stricter breeding" (1927, p. 376).

But, most importantly, Spearman simply didn't seem to take much interest in the subject of hereditary differences among peoples. While the issue swirled about him and buried his profession in printer's ink, and while he himself had supplied a basic argument for the hereditarian school, the inventor of *g* stood aside in apparent apathy. He had studied factor analysis because he wanted to understand the structure of the human brain, not as a guide to measuring differences between groups, or even among individuals. Spearman may have been a reluctant courtier, but the politically potent union of IQ and factor analysis into a hereditarian theory of intelligence was engineered by Spearman's successor in the chair of psychology at University College—Cyril Burt. Spearman may have cared little, but the innate character of intelligence was the idée fixe of Sir Cyril's life.

Cyril Burt and the hereditarian synthesis

The source of Burt's uncompromising hereditarianism

Cyril Burt published his first paper in 1909. In it, he argued that intelligence is innate and that differences between social classes are largely products of heredity; he also cited Spearman's *g* as primary support. Burt's last paper in a major journal appeared posthumously in 1972. It sang the very same tune: intelligence is innate and the existence of Spearman's *g* proves it. For all his more dubious qualities, Cyril Burt certainly had staying power. The 1972 paper proclaims:

> The two main conclusions we have reached seem clear and beyond all question. The hypothesis of a general factor entering into every type of cognitive process, tentatively suggested by speculations derived from neurology and biology, is fully borne out by the statistical evidence; and the contention that differences in this general factor depend largely on the individual's genetic constitution appears incontestable. The concept of an innate, general, cognitive ability, which follows from these two assumptions, though admittedly a sheer abstraction, is thus wholly consistent with the empirical facts (1972, p. 188).

Only the intensity of Sir Cyril's adjectives had changed. In 1912 he had termed this argument "conclusive"; by 1972 it had become "incontestable."

Factor analysis lay at the core of Burt's definition of intelligence as i.g.c. (innate, general, cognitive) ability. In his major work on factor analysis (1940, p. 216), Burt developed his characteristic use of Spearman's thesis. Factor analysis shows that "a *general* factor enters into all *cognitive* processes," and "this general factor appears to be largely, if not wholly, inherited or *innate*"—again, i.g.c. ability. Three years earlier (1937, pp. 10–11) he had tied *g* to an ineluctable heredity even more graphically:

> This general intellectual factor, central and all-pervading, shows a further characteristic, also disclosed by testing and statistics. It appears to be inherited, or at least inborn. Neither knowledge nor practice, neither interest nor industry, will avail to increase it.

Others, including Spearman himself, had drawn the link between *g* and heredity. Yet no one but Sir Cyril ever pursued it with such stubborn, almost obsessive gusto: and no one else wielded it as such an effective political tool. The combination of hereditarian bias with a reification of intelligence as a single, measurable entity defined Burt's unyielding position.

I have discussed the roots of the second component: intelligence as a reified factor. But where did the first component—rigid hereditarianism—arise in Burt's view of life? It did not flow logically from factor analysis itself, for it cannot (see pp. 250–252). I will not attempt to answer this question by referring either to Burt's psyche or his times (though Hearnshaw, 1979, has made some suggestions). But I will demonstrate that Burt's hereditarian argument had no foundation in his empirical work (either honest or fraudulent), and that it represented an a priori bias imposed upon the studies that supposedly proved it. It also acted, through Burt's zealous pursuit of his idée fixe, as a distorter of judgment and finally as an incitement to fraud.*

BURT'S INITIAL "PROOF" OF INNATENESS

Throughout his long career, Burt continually cited his first paper of 1909 as a proof that intelligence is innate. Yet the study falters both on a flaw of logic (circular reasoning) and on the remarkably scant and superficial character of the data themselves. This publication proves only one thing about intelligence—that Burt began his study with an a priori conviction of its innateness, and reasoned back in a vicious circle to his initial belief. The "evidence"—what there was of it—served only as selective window dressing.

At the outset of his 1909 paper, Burt set three goals for himself. The first two reflect the influence of Spearman's pioneering work in factor analysis ("can general intelligence be detected and measured"; "can its nature be isolated and its meaning analyzed"). The third represents Burt's peculiar concern: "Is its development predominantly determined by environmental influence and individual acquisition, or is it rather dependent upon the inheritance of a racial character or family trait" (1909, p. 96).

Not only does Burt proclaim this third question "in many ways

* Of Burt's belief in the innateness of intelligence, Hearnshaw writes (1979, p. 49): "It was for him almost an article of faith, which he was prepared to defend against all opposition, rather than a tentative hypothesis to be refuted, if possible, by empirical tests. It is hard not to feel that almost from the first Burt showed an excessive assurance in the finality and correctness of his conclusions."

the most important of all," but he also gives away his answer in stating why we should be so concerned. Its importance rests upon:

> . . . the growing belief that innate characters of the family are more potent in evolution than the acquired characters of the individual, the gradual apprehension that unsupplemented humanitarianism and philanthropy may be suspending the natural elimination of the unfit stocks—these features of contemporary sociology make the question whether ability is inherited one of fundamental moment (1909, p. 169).

Burt selected forty-three boys from two Oxford schools, thirty sons of small tradesmen from an elementary school and thirteen upper-class boys from preparatory school. In this "experimental demonstration that intelligence is hereditary" (1909, p. 179), with its ludicrously small sample, Burt administered twelve tests of "mental functions of varying degrees of complexity" to each boy. (Most of these tests were not directly cognitive in the usual sense, but more like the older Galtonian tests of physiology—attention, memory, sensory discrimination, and reaction time). Burt then obtained "careful empirical estimates of intelligence" for each boy. This he did not by rigorous Binet testing, but by asking "expert" observers to rank the boys in order of their intelligence independent of mere school learning. He obtained these rankings from the headmasters of the schools, from teachers, and from "two competent and impartial boys" included in the study. Writing in the triumphant days of British colonialism and derring-do, Burt instructed his two boys on the meaning of intelligence:

> Supposing you had to choose a leader for an expedition into an unknown country, which of these 30 boys would you select as the most intelligent? Failing him, which next? (1909, p. 106)

Burt then searched for correlations between performance on the twelve tests and the rankings produced by his expert witnesses. He found that five tests had correlation coefficients with intelligence above 0.5, and that poorest correlations involved tests of "lower senses—touch and weight," while the best correlations included tests of clearer cognitive import. Convinced that the twelve tests measured intelligence, Burt then considered the scores themselves. He found that the upper-class boys performed better than the lower-middle-class boys in all tests save those involving weight and touch. The upper-class boys must therefore be smarter.

But is the superior smartness of upper-class boys innate or

acquired as a function of advantages in home and schooling? Burt gave four arguments for discounting environment:

1. The environment of lower-middle-class boys cannot be poor enough to make a difference since their parents can afford the ninepence a week required to attend school: "Now in the case of the lowest social classes, general inferiority at mental tests might be attributable to unfortunate environmental and post-natal influences. . . . But such conditions could not be suspected with the boys who, at a fee of 9d a week, attended the Central Elementary School" (1909, p. 173). In other words, environment can't make a difference until it reduces a child to near starvation.

2. The "educative influences of home and social life" seem small. In making this admittedly subjective assessment, Burt appealed to a fine intuition honed by years of gut-level experience. "Here, however, one must confess, such speculative arguments can convey little conviction to those who have not witnessed the actual manner of the respective boys."

3. The character of the tests themselves precludes much environmental influence. As tests of sensation and motor performance, they do not involve "an appreciable degree of acquired skill or knowledge. . . . There is reason, therefore, to believe that the differences revealed are mainly innate" (1909, p. 180).

4. A retesting of the boys eighteen months later, after several had entered professions or new schools, produced no important readjustment of ranks. (Did it ever occur to Burt that environment might have its primary influence in early life, and not only in immediate situations?)

The problem with all these points, and with the design of the entire study, is a patent circularity in argument. Burt's claim rested upon correlations between test performances and a ranking of intelligence compiled by "impartial" observers. (Arguments about the "character" of the tests themselves are secondary, for they would count for nothing in Burt's design if the tests did not correlate with independent assessments of intelligence.) We must know what the subjective rankings mean in order to interpret the correlations and make any use of the tests themselves. For if the rankings of teachers, headmasters, and colleagues, however sincerely attempted, record the advantages of upbringing more than the differential blessings of genetics, then the ranks are primarily a record of environment, and the test scores may provide just another (and more imperfect) measure of the same thing. Burt used the correlation between two criteria as evidence for heredity without ever establishing that either criterion measured his favored property.

In any case, all these arguments for heredity are indirect. Burt also claimed, as his final proof, a direct test of inheritance: the boys' measured intelligence correlated with that of their parents:

> Wherever a process is correlated with intelligence, these children of superior parentage resemble their parents in being themselves superior. . . . Proficiency at such tests does not depend upon opportunity or training, but upon some quality innate. The resemblance in degree of intelligence between the boys and their parents must, therefore, be due to inheritance. We thus have an experimental demonstration that intelligence is hereditary (1909, p. 181).

But how did Burt measure parental intelligence? The answer, remarkable even from Burt's point of view, is that he didn't: he merely assumed it from profession and social standing. Intellectual, upper-class parents must be innately smarter than tradesmen. But the study was designed to assess whether or not performance on tests reflects inborn qualities or the advantages of social standing. One cannot, therefore, turn around and infer intelligence directly from social standing.

We know that Burt's later studies of inheritance were fraudulent. Yet his early and honest work is riddled with flaws so fundamental that they stand in scarcely better light. As in the 1909 study, Burt continually argued for innateness by citing correlations in intelligence between parents and offspring. And he continually assessed parental intelligence by social standing, not by actual tests.

For example, after completing the Oxford study, Burt began a more extensive program of testing in Liverpool. He cited high correlations between parents and offspring as a major argument for innate intelligence, but never provided parental scores. Fifty years later, L. S. Penrose read Burt's old work, noted the absent data, and asked Burt how he had measured parental intelligence. The old man replied (in Hearnshaw, 1979, p. 29):

> The intelligence of the parents was assessed primarily on the basis of their actual jobs, checked by personal interviews; about a fifth were also tested to standardize the impressionistic assessments.

Hearnshaw comments (1979, p. 30): "Inadequate reporting and incautious conclusions mark this first incursion of Burt into the genetic field. We have here, right at the beginning of his career, the seeds of later troubles."

Even when Burt did test subjects, he rarely reported the actual scores as measured, but "adjusted" them according to his own assessment of their failure to measure true intelligence as he and other experts subjectively judged it. He admitted in a major work (1921, p. 280):

> I did not take my test results just as they stood. They were carefully discussed with teachers, and freely corrected whenever it seemed likely that the teacher's view of the relative merits of his own pupils gave a better estimate than the crude test marks.

Such a procedure is not without its commendable intent. It does admit the inability of a mere number, calculated during a short series of tests, to capture such a subtle notion as intelligence. It does grant to teachers and others with extensive personal knowledge the opportunity to record their good judgment. But it surely makes a mockery of any claim that a specific hypothesis is under objective and rigorous test. For if one believes beforehand that well-bred children are innately intelligent, then in what direction will the scores be adjusted?*

Despite his minuscule sample, his illogical arguments, and his dubious procedures, Burt closed his 1909 paper with a statement of personal triumph (p. 176):

> Parental intelligence, therefore, may be inherited, individual intelligence measured, and general intelligence analyzed; and they can be analyzed, measured and inherited to a degree which few psychologists have hitherto legitimately ventured to maintain.

When Burt recycled these data in a 1912 paper for the *Eugenics Review,* he added additional "proof" with even smaller samples. He

*Sometimes, Burt descended even further into circular illogic and claimed that tests must measure innate intelligence because the testers constructed them to do so: "Indeed from Binet onwards practically all the investigators who have attempted to construct 'intelligence tests' have been primarily searching for some measure of *inborn* capacity, as distinct from acquired knowledge or skill. With such an interpretation it obviously becomes foolish to inquire how far 'intelligence' is due to environment and how far it is due to innate constitution: the very definition begs and settles the question" (1943, p. 88).

discussed Alfred Binet's two daughters, noted that their father had been disinclined to connect physical signs with mental prowess, and pointed out that the blond, blue-eyed, large-headed daughter of Teutonic appearance was objective and forthright, while the darker daughter tended to be impractical and sentimental. *Touché.*

Burt was no fool. I confess that I began reading him with the impression, nurtured by spectacular press reports of his fraudulent work, that he was simply a devious and foxy charlatan. To be sure, that he became and for complex reasons (see pp. 234–239). But as I read, I gained respect for Burt's enormous erudition, for his remarkable sensitivity in most areas, and for the subtlety and complexity of his reasoning; I ended up liking most things about him in spite of myself. And yet, this assessment makes the extraordinary weakness of his reasoning about the innateness of intelligence all the more puzzling. If he had simply been a fool, then foolish arguments would denote consistency of character.

My dictionary defines an *idée fixe,* or fixed idea, as "a persistent or obsessing idea, often delusional, from which a person cannot escape." The innateness of intelligence was Burt's idée fixe. When he turned his intellectual skills to other areas, he reasoned well, subtly, and often with great insight. When he considered the innateness of intelligence, blinders descended and his rational thinking evaporated before the hereditarian dogma that won his fame and eventually sealed his intellectual doom. It may be remarkable that Burt could operate with such a duality in styles of reasoning. But I find it much more remarkable that so many others believed Burt's statements about intelligence when his arguments and data, all readily available in popular publications, contained such patent errors and specious claims. What does this teach us about shared dogma masquerading as objectivity?

LATER ARGUMENTS

Perhaps I have been unfair in choosing Burt's earliest work for criticism. Perhaps the foolishness of youth soon yielded to mature wisdom and caution. Not at all; Burt was nothing if not ontogenetically consistent. The argument of 1909 never changed, never gained subtlety, and ended with manufactured support. The innateness of intelligence continued to function as dogma. Consider the primary argument of Burt's most famous book, *The Back-*

ward Child (1937), written at the height of his powers and before his descent into conscious fraud.

Backwardness, Burt notes, is defined by achievement in school, not by tests of intelligence: backward children are more than a year behind in their schoolwork. Burt argues that environmental effects, if at all important, should have most impact upon children in this category (those much further behind in school are more clearly genetically impaired). Burt therefore undertook a statistical study of environment by correlating the percentage of backward children with measures of poverty in the boroughs of London. He calculated an impressive array of strong correlations: 0.73 with percentage of people below the poverty line, 0.89 with overcrowding, 0.68 with unemployment, and 0.93 with juvenile mortality. These data seem to provide a prima-facie case for a dominant environmental influence upon backwardness, but Burt demurs. There is another possibility. Perhaps the innately poorest stocks create and then gravitate to the worst boroughs, and degree of poverty is merely an imperfect measure of genetic worthlessness.

Burt, guided by his idée fixe, opted for innate stupidity as the primary cause of poverty (1937, p. 105). He invoked IQ testing as his major argument. Most backward children score 1 to 2 standard deviations below the mean (70–85), within a range technically designated as "dull." Since IQ records innate intelligence, most backward children perform poorly in school because they are dull, not (or only indirectly) because they are poor. Again, Burt rides his circle. He wishes to prove that deficiency of innate intelligence is the major cause of poor performance in school. He knows full well that the link between IQ score and innateness is an unresolved issue in intense debates about the meaning of IQ—and he admits in many places that the Stanford-Binet test is, at best, only an imperfect measure of innateness (e.g., 1921, p. 90). Yet, using the test scores as a guide, he concludes:

> In well over half the cases, the backwardness seems due chiefly to intrinsic mental factors; here, therefore, it is primary, innate, and to that extent beyond all hope of cure (1937, p. 110).

Consider Burt's curious definition of innate in this statement. An innate character, as inborn and, in Burt's usage, inherited, forms part of an organism's biological constitution. But the demonstration that a trait represents nature unaffected by nurture does not guarantee its ineluctable state. Burt inherited poor vision. No doctor ever rebuilt his eyes to an engineer's paradigm of normal design, but Burt wore eyeglasses and the only clouding of his vision was conceptual.

The Backward Child also abounds in tangential statements that record Burt's hereditarian biases. He writes about an environmental handicap—recurrent catarrh among the poor—and discusses hereditary susceptibility (quite plausible) with an arresting quip for graphic emphasis:

> . . . exceptionally prevalent in those whose faces are marked by developmental defects—by the round receding forehead, the protruding muzzle, the short and upturned nose, the thickened lips, which combine to give to the slum child's profile a negroid or almost simian outline. . . . "Apes that are hardly anthropoid" was the comment of one headmaster, who liked to sum up his cases in a phrase (1937, p. 186).

He wonders about the intellectual achievement of Jews and attributes it, in part, to inherited myopia that keeps them off the playing fields and adapts them for poring over account books.

> Before the invention of spectacles, the Jew whose living depended upon his ability to keep accounts and read them, would have been incapacitated by the age of 50, had he possessed the usual tendency to hypermetropia: on the other hand (as I can personally testify) the myope . . . can dispense with glasses for near work without much loss of efficiency (1937, p. 219).

BURT'S BLINDNESS

The blinding power of Burt's hereditarian biases can best be appreciated by studying his approach to subjects other than intelligence. For here he consistently showed a commendable caution. He recognized the complexity of causation and the subtle influence that environment can exert. He railed against simplistic assumptions and withheld judgment pending further evidence. Yet as soon as Burt returned to his favorite subject of intelligence, the blinders descended and the hereditarian catechism came forward again.

Burt wrote with power and sensitivity about the debilitating effects of poor environments. He noted that 23 percent of the cockney youth he interviewed had never seen a field or a patch of grass, not "even in a Council park," 64 percent had never seen a train, and 98 percent had never seen the sea. The following passage displays a measure of paternalistic condescension and stereotyping, but it also presents a powerful image of poverty in working-class homes, and its intellectual effect upon children (1937, p. 127).

> His mother and father know astonishingly little of any life except their own, and have neither the time nor the leisure, neither the ability nor the disposition, to impart what little they know. The mother's conversation may be chiefly limited to the topics of cleaning, cooking, and scolding. The father, when not at work, may spend most of his time "round the corner" refreshing a worn-out body, or sitting by the fire with cap on and coat off, sucking his pipe in gloomy silence. The vocabulary that the child absorbs is restricted to a few hundred words, most of them inaccurate, uncouth, or mispronounced, and the rest unfit for reproduction in the schoolroom. In the home itself there is no literature that deserves the title; and the child's whole universe is closed in and circumscribed by walls of brick and a pall of smoke. From one end of the year to the other, he may go no farther than the nearest shops or the neighborhood recreation ground. The country or the seaside are mere words to him, dimly suggesting some place to which cripples are sent after an accident, visualized perhaps in terms of some photographic "souvenir from Southend" or some pictorial "memento from Margate," all framed in shells, brought back by his parents on a bank-holiday trip a few weeks after their wedding.

Burt appended this comment from a "burly bus conductor" to his description: "Book learning isn't for kids that'll have to earn their bread. It's only for them as likes to give themselves the hairs of the 'ighbrow."

Burt could apply what he understood so well to subjects other than intelligence. Consider his views on juvenile delinquency and left-handedness. Burt wrote extensively on the cause of delinquency and attributed it to complex interactions between children and their environment: "The problem never lies in the 'problem child' alone: it lies always in the relations between that child and his environment" (1940, p. 243). If poor behavioral performance merits such an assessment, why not say the same about poor intellectual performance? One might suspect that Burt relied again upon test scores, arguing that delinquents tested well and could not be misbehaving as a result of innate stupidity. But, in fact, delinquents often tested as badly as poor children regarded by Burt as innately deficient in intelligence. Yet Burt recognized that IQ scores of delinquents may not reflect inherited ability because they rebel against taking the tests:

> For what to them must seem nothing but a resuscitated school examination, delinquents, as a rule, feel little inclination and much distaste. From the outset they assume they are more likely to fail than succeed, more likely to be reproached than commended. . . . Unless, indeed, to circumvent their suspicion and secure their good-will special manoeuvers be tactfully tried, their apparent prowess with all such tests will fall much below their veritable powers. . . . In the causation of juvenile delinquency . . . the share contributed by mental defect has unquestionably been magnified by those who, trusting so exclusively to the Binet-Simon scale, have ignored the factors which depreciate its results (1921, pp. 189–190).

But why not say that poverty often entails a similar disinclination and sense of defeat?

Burt (1937, p. 270) regarded left-handedness as the "motor disability . . . which interferes most widely with the ordinary tasks of the classroom." As chief psychologist of the London schools, he therefore devoted much study to its cause. Unburdened by a priori conviction in this case, he devised and attempted to test a wide range of potential environmental influences. He studied medieval and Renaissance paintings to determine if Mary usually carried the infant Jesus on her right hip. If so, babies would wrap their left arms about their mother's neck, leaving their right hand free for more dextrous (literally right-handed) motion. He wondered if greater frequency of right-handedness might record the asymmetry of internal organs and the need for protection imposed by our habits. If heart and stomach lie to the left of the midline, then a warrior or worker would naturally turn his left side away from potential danger, "trust to the more solid support of the right side of the trunk, and so use his right hand and arm for wielding heavy instruments and weapons" (1937, p. 270). In the end, Burt opted for caution and concluded that he could not tell:

> I should in the last resort contend that probably all forms of left-handedness are only indirectly hereditary: postnatal influence seems always to enter in. . . . I must accordingly repeat that, here as elsewhere in psychology, our present knowledge is far too meager to allow us to declare with any assurance what is inborn and what is not (1937, pp. 303–304).

Substitute "intelligence" for "left-handedness" and the statement is a model of judicious inference. In fact, left-handedness is more clearly an entity than intelligence, and probably more subject to definite and specifiable hereditary influence. Yet here, where his case for innateness was better, Burt tested all the environmental influences—some rather farfetched—that he could devise, and finally declared the subject too complex for resolution.

BURT'S POLITICAL USE OF INNATENESS

Burt extended his belief in the innateness of individual intelligence to only one aspect of averate differences between groups. He did not feel (1912) that races varied much in inherited intelligence, and he argued (1921, p. 197) that the different behaviors of boys and girls can be traced largely to parental treatment. But differences in social class, the wit of the successful and dullness of the poor, are reflections of inherited ability. If race is America's primary social problem, then class has been Britain's corresponding concern.

In his watershed* paper (1943) on "ability and income," Burt concludes that "the wide inequality in personal income is largely, though not entirely, an indirect effect of the wide inequality in innate intelligence." The data "do not support the view (still held by many educational and social reformers) that the apparent inequality in intelligence of children and adults is in the main an indirect consequence of inequality in economic conditions" (1943, p. 141).

Burt often denied that he wished to limit opportunities for achievement by regarding tests as measures of innate intelligence. He argued, on the contrary, that tests could identify those few individuals in the lower classes whose high innate intelligence would not otherwise be recognized under a veneer of environmental disadvantage. For "among nations, success in the struggle for survival is bound to depend more and more on the achievements of a small handful of individuals who are endowed by nature with outstanding gifts of ability and character" (1959, p. 31). These people must be identified and nurtured to compensate for "the comparative ineptitude of the general public" (1959, p. 31). They must be encouraged and rewarded, for the rise and fall of a nation does not depend upon genes peculiar to an entire race, but upon

*Hearnshaw (1979) suspects that this paper marks Burt's first use of fraudulent data.

"changes in the relative fertility of its leading members or its leading classes" (1962, p. 49).

Tests may have been the vehicle by which a few children escaped from the strictures of a fairly inflexible class structure. But what was their effect on the vast majority of lower-class children whom Burt unfairly branded as unable, by inheritance, ever to develop much intelligence—and therefore undeserving, by reason, of higher social standing?

> Any recent attempt to base our educational policy for the future on the assumption that there are no real differences, or at any rate no important differences, between the average intelligence of the different social classes, is not only bound to fail; it is likely to be fraught with disastrous consequences for the welfare of the nation as a whole, and at the same time to result in needless disappointments for the pupils concerned. The facts of genetic inequality, whether or not they conform to our personal wishes and ideals, are something that we cannot escape (1959, p. 28). . . . A definite limit to what children can achieve is inexorably set by the limitations of their innate capacity (1969).

Burt's extension of Spearman's theory

Cyril Burt may be known best to the public as a hereditarian in the field of mental testing, but his reputation as a theoretical psychologist rested primarily upon his work in factor analysis. He did not invent the technique, as he later claimed; but he was Spearman's successor, both literally and figuratively, and he became the leading British factorist of his generation.

Burt's genuine achievements in factor analysis were substantial. His complex and densely reasoned book on the subject (1940) was the crowning achievement of Spearman's school. Burt wrote that it "may prove to be a more lasting contribution to psychology than anything else I have yet written" (letter to his sister quoted in Hearnshaw, 1979, p. 154). Burt also pioneered (though he did not invent) two important extensions of Spearman's approach—an inverted technique (discussed on pp. 292–293) that Burt called "correlation between persons" (now known to aficionados as "Q-mode factor analysis"), and an expansion of Spearman's two-factor theory to add "group factors" at a level between *g* and *s*.

Burt toed Spearman's line in his first paper of 1909. Spearman had insisted that each test recorded only two properties of mind—

a general factor common to all tests and a specific factor peculiar to that test alone. He denied that clusters of tests showed any significant tendency to form "group factors" between his two levels—that is, he found no evidence for the "faculties" of an older psychology, no clusters representing verbal, spatial, or arithmetic ability, for example. In his 1909 paper, Burt did note a "discernible, but small" tendency for grouping in allied tests. But he proclaimed it weak enough to ignore ("vanishingly minute" in his words), and argued that his results "confirm and extend" Spearman's theory.

But Burt, unlike Spearman, was a practitioner of testing (responsible for all of London's schools). Further studies in factor analysis continued to distinguish group factors, though they were always subsidiary to *g*. As a practical matter for guidance of pupils, Burt realized that he could not ignore the group factors. With a purely Spearmanian approach, what could a pupil be told except that he was generally smart or dumb? Pupils had to be guided toward professions by identifying strengths and weaknesses in more specific areas.

By the time Burt did his major work in factor analysis, Spearman's cumbersome method of tetrad differences had been replaced by the principal components approach outlined on pp. 245–250. Burt identified group factors by studying the projection of individual tests upon the second and subsequent principal components. Consider Fig. 6.6: In a matrix of positive correlation coefficients, vectors representing individual tests are all clustered together. The first principal component, Spearman's *g* runs through the middle of the cluster and resolves more information than any other axis could. Burt recognized that no consistent patterns would be found on subsequent axes if Spearman's two-factor theory held—for the vectors would not form subclusters if their only common variance had already been accounted for by *g*. But if the vectors form subclusters representing more specialized abilities, then the first principal component must run *between* the subclusters if it is to be the best average fit to all vectors. Since the second principal component is perpendicular to the first, some subclusters must project positively upon it and others negatively (as Fig. 6.6 shows with its negative projections for verbal tests and positive projections for arithmetic tests). Burt called these axes *bipolar factors,* because they included clusters of positive and negative projections. He identified as *group factors* the clusters of positive and negative projections themselves.

Burt's identification of group factors may seem, superficially, to challenge Spearman's theory, but in fact it provided an extension and improvement that Spearman eventually welcomed. The essence of Spearman's claim is the primacy of *g*, and the subordination of all other determinants of intelligence to it. Burt's identification of group factors preserved this notion of hierarchy, and extended it by adding another level between *g* and *s*. In fact, Burt's treatment of group factors as a level in a hierarchy subordinate to *g* saved Spearman's theory from the data that seemed to threaten it. Spearman originally denied group factors, but evidence for them continued to accumulate. Many factorists began to view this evidence as a denigration of *g* and as a wedge for toppling Spearman's entire edifice. Burt strengthened the building, preserved the preeminent role of *g*, and extended Spearman's theory by enumerating further levels subordinate to *g*. The factors, Burt wrote (1949, p. 199), are "organized on what may be called a hierarchical basis. . . . There is first a comprehensive general factor, covering all cognitive activities; next a comparatively small number of broad group factors, covering different abilities classified according to their form or content. . . . The whole series appears to be arranged on successive levels, the factors on the lowest level being the most specific and the most numerous of all."

Spearman had advocated a two-factor theory; Burt proclaimed a four-factor theory: the *general* factor or Spearman's *g*, the particular or *group* factors that he had identified, the *specific* factors or Spearman's *s* (attributes of a single trait measured on all occasions), and what Burt called *accidental* factors, or attributes of a single trait measured only on a single occasion.* Burt had synthesized all perspectives. In Spearman's terms, his theory was monarchic in recognizing the domination of *g*, oligarchic in its identification of group factors, and anarchic in recognizing *s*-factors for each test. But Burt's scheme was no compromise; it was Spearman's hierarchical theory with yet another level subordinate to *g*.

*This accidental variance, representing peculiarities of particular testing situations, forms part of what statisticians call "measurement error." It is important to quantify, for it may form a basic level of comparison for the identification of causes in a family of techniques called the "analysis of variance." But it represents the peculiarity of an occasion, not a quality either of a test or a testee.

Moreover, Burt accepted and greatly elaborated Spearman's views on the differential innateness of levels. Spearman had regarded *g* as inherited, *s* as a function of training. Burt agreed, but promoted the influence of education to his group factors as well. He retained the distinction between an inherited and ineluctable *g*, and a set of more specialized abilities amenable to improvement by education:

> Although defect in general intelligence inevitably places a definite limit to educational progress, defect in special intellectual abilities rarely does so (1937, p. 537).

Burt also declared, with his usual intensity and persistence, that the primary importance of factor analysis lay in its capacity for identifying inherited, permanent qualities:

> From the very outset of my educational work it has seemed essential, not merely to show that a general factor underlies the cognitive group of mental activities, but also that this general factor (or some important component of it) is innate or permanent (1940, p. 57).
>
> The search for factors thus becomes, to a great extent, an attempt to discover inborn potentialities, such as will permanently aid or limit the individual's behavior later on (1940, p. 230).

Burt on the reification of factors

Burt's view on reification, as Hearnshaw has noted with frustration (1979, p. 166), are inconsistent and even contradictory (sometimes within the same publication).* Often, Burt branded reification of factors as a temptation to be avoided:

> No doubt, this causal language, which we all to some extent favor, arises partly from the irrepressible disposition of the human mind to reify and even to personify whatever it can—to picture inferred reasons as realities and to endow those realities with an active force (1940, p. 66).

*Other scholars often complained of Burt's tendency to obfuscate, temporize, and argue both sides as his own when treating difficult and controversial issues. D. F. Vincent wrote of his correspondence with Burt about the history of factor analysis (in Hearnshaw, 1979, pp. 177–178): "I should not get a simple answer to a simple question. I should get half a dozen foolscap sheets of typescript, all very polite and very cordial, raising half a dozen subsidiary issues in which I was not particularly interested, and to which out of politeness I should have to reply . . . I should then get more foolscap pages of typescript raising more extraneous issues. . . . After the first letter my problem has been how to terminate the correspondence without being discourteous."

He spoke with eloquence about this error of thought:

> The ordinary mind loves to reduce patterns to single atomlike existents—to treat memory as an elementary faculty lodged in a phrenological organ, to squeeze all consciousness into the pineal gland, to call a dozen different complaints rheumatic and regard them all as the effect of a specific germ, to declare that strength resides in the hair or in the blood, to treat beauty as an elementary quality that can be laid on like so much varnish. But the whole trend of current science is to seek its unifying principles, not in simple unitary causes, but in the system or structural pattern as such (1940, p. 237).

And he explicitly denied that factors were things in the head (1937, p. 459):

> The "factors," in short, are to be regarded as convenient mathematical abstractions, not as concrete mental "faculties," lodged in separate "organs" of the brain.

What could be more clearly stated?

Yet in a biographical comment, Burt (1961, p. 53) centered his argument with Spearman not on the issue of whether or not factors should be reified, but rather *how* they should be reified: "Spearman himself identified the general factor with 'cerebral energy.' I identified it with the general structure of the brain." In the same article, he provided more details of suspected physical locations for entities identified by mathematical factors. Group factors, he argues, are definite areas of the cerebral cortex (1961, p. 57), while the general factor represents the amount and complexity of cortical tissue: "It is this general character of the individual's brain-tissue—viz., the general degree of systematic complexity in the neuronal architecture—that seems to me to represent the general factor, and account for the high positive correlations obtained between various cognitive tests" (1961, pp. 57–58; see also 1959, p. 106).*

*One might resolve this apparent contradiction by arguing that Burt refused to reify on the basis of mathematical evidence alone (in 1940), but did so later when independent neurological information confirmed the existence of structures in the brain that could be identified with factors. It is true that Burt advanced some neurological arguments (1961, p. 57, for example) in comparing the brains of normal individuals and "low grade defectives." But these arguments are sporadic, perfunctory, and peripheral. Burt repeated them virtually verbatim, in publication after publication, without citing sources or providing any specific reason for allying mathematical factors with cortical properties.

Lest one be tempted to regard these later statements as a shift in belief from the caution of a scholar in 1940 to the poor judgment of a man mired in the frauds of his later years, I note that Burt presented the same arguments for reification in 1940, right alongside the warnings against it:

> Now, although I do not identify the general factor *g* with any form of energy, I should be ready to grant it quite as much "real existence" as physical energy can justifiably claim (1940, p. 214). Intelligence I regard not indeed as designating a special form of energy, but rather as specifying certain individual differences in the structure of the central nervous system—differences whose concrete nature could be described in histological terms (1940, pp. 216–217).

Burt even went so far as to suggest that the all-or-none character of neural discharge "supports the demand for an ultimate analysis into independent or 'orthogonal' factors" (1940, p. 222).

But perhaps the best indication of Burt's hope for reification lies in the very title he chose for his major book of 1940. He called it *The Factors of the Mind.*

Burt followed Spearman in trying to find a physical location in the brain for mathematical factors extracted from the correlation matrix of mental tests. But Burt also went further, and established himself as a reifier in a domain that Spearman himself would never have dared to enter. Burt could not be satisfied with something so vulgar and material as a bit of neural tissue for the residence of factors. He had a wider vision that evoked the spirit of Plato himself. Material objects on earth are immediate and imperfect representations of higher essences in an ideal world beyond our ken.

Burt subjected many kinds of data to factor analysis during his long career. His interpretations of factors display a Platonic belief in a higher reality, embodied imperfectly by material objects, but discernible in them through an idealization of their essential, underlying properties on principal component factors. He analyzed a suite of emotional traits (1940, pp. 406–408) and identified his first principal component as a factor of "general emotionality." (He also found two bipolar factors for extrovert-introvert and euphoric-sorrowful.) He discovered "a general paranormal factor" in a study of ESP data (in Hearnshaw, 1979, p. 222). He analyzed human anatomy and interpreted the first principal component as an ideal type for humanity (1940, p. 113).

One needn't, from these examples, infer Burt's belief in a literal, higher reality: perhaps he thought of these idealized general factors as mere principles of classification to aid human understanding. But, in a factor analysis of aesthetic judgment, Burt explicitly expressed his conviction that real standards of beauty exist, independent of the presence of human beings to appreciate them. Burt selected fifty postcards with illustrations ranging from the great masters down to "the crudest and most flashy birthday card that I could find at a paper shop in the slums." He asked a group of subjects to rank the cards in order of beauty and performed a factor analysis of correlations among the ranks. Again, he discerned an underlying general factor on the first principal component, declared it to be a universal standard of beauty, and expressed a personal contempt for Victorian ceremonial statuary in identifying this higher reality:

> We see beauty because it is there to be seen. . . . I am tempted to contend that aesthetic relations, like logical relations, have an independent, objective existence: the Venus of Milo would remain more lovely than Queen Victoria's statue in the Mall, the Taj Mahal than the Albert Memorial, though every man and woman in the world were killed by a passing comet's gas.

In analyses of intelligence, Burt often claimed (1939, 1940, 1949, for example) that each level of his hierarchical, four-factor theory corresponded with a recognized category in "the traditional logic of classes" (1939, p. 85)—the general factor to the *genus,* group factors to *species,* specific factors to the *proprium,* and accidental factors to the *accidens.* He seemed to regard these categories as more than conveniences for human ordering of the world's complexity, but as necessary ways of parsing a hierarchically structured reality.

Burt certainly believed in realms of existence beyond the material reality of everyday objects. He accepted much of the data of parapsychology and postulated an oversoul or psychon—"a kind of group mind formed by the subconscious telepathic interaction of the minds of certain persons now living, together perhaps with the psychic reservoir out of which the minds of individuals now deceased were formed, and into which they were reabsorbed on the death of their bodies" (Burt quoted in Hearnshaw, 1979, p. 225). In this higher realm of psychic reality, the "factors of the mind" may have real existence as modes of truly universal thought.

Burt managed to espouse three contradictory views about the nature of factors: mathematical abstractions for human convenience; real entities lodged in physical properties of the brain; and real categories of thought in a higher, hierarchically organized realm of psychic reality. Spearman had not been very daring as a reifier; he never ventured beyond the Aristotelian urge for locating idealized abstractions within physical bodies themselves. Burt, at least in part, soared beyond into a Platonic realm above and beyond physical bodies. In this sense, Burt was the boldest, and literally most extensive, reifier of them all.

Burt and the political uses of g

Factor analysis is usually performed on the correlation matrix of tests. Burt pioneered an "inverted" form of factor analysis, mathematically equivalent to the usual style, but based on correlation between persons rather than tests. If each vector in the usual style (technically called R-mode analysis) represents the scores of several people on a single test, then each vector in Burt's inverted style (called Q-mode analysis) reflects the results of several tests for a single person. In other words, each vector now represents a person rather than a test, and the correlation between vectors measures the degree of relationship between individuals.

Why did Burt go to such lengths to develop a technique mathematically equivalent to the usual form, and generally more cumbersome and expensive to apply (since an experimental design almost always includes more people than tests)? The answer lies in Burt's uncommon focus of interest. Spearman, and most other factorists, wished to learn about the nature of thought or the structure of mind by studying correlations between tests measuring different aspects of mental functioning. Cyril Burt, as official psychologist of the London County Council (1913–1932), was interested in ranking pupils. Burt wrote in an autobiographical statement (1961, p. 56): "[Sir Godfrey] Thomson was interested primarily in the description of the *abilities* tested and in the differences between those abilities; I was interested rather in the *persons* tested and in the differences between them" (Burt's italics).

Comparison, for Burt, was no abstract issue. He wished to assess pupils in his own characteristic way, based upon two guiding principles: first (the theme of this chapter) that general intelligence is a single, measurable entity (Spearman's *g*); second (Burt's own idée fixe) that a person's general intelligence is almost entirely innate and unchangeable. Thus, Burt sought the relationship among persons in a *unilinear ranking of inherited mental worth.* He used factor analysis to validate this single scale and to plant people upon it. "The very object of the factor-analysis," he wrote (1940, p. 136), "is to deduce from an empirical set of test measurements a single figure for each single individual." Burt sought (1940, p. 176) "one ideal order, acting as a general factor, common to every examiner and to every examinee, predominating over, though no doubt disturbed by, other irrelevant influences."

Burt's vision of a single ranking based on inherited ability fueled the major political triumph in Britain of hereditarian theories of mental testing. If the Immigration Restriction Act of 1924 signalled the chief victory of American hereditarians in psychology, then the so-called examination at 11+ awarded their British counterparts a triumph of equal impact. Under this system for streaming children into different secondary schools, pupils took an extensive examination at age ten or eleven. As a result of these tests, largely an attempt to assess Spearman's *g* for each child, 20 percent were sent to "grammar" schools where they might prepare for entry to a university, while 80 percent were relegated to technical or "secondary modern" schools and regarded as unfit for higher education.

Cyril Burt defended this separation as a wise step for "warding off the ultimate decline and fall that has overtaken each of the great civilizations of the past" (1959, p. 117):

> It is essential in the interests alike of the children themselves and of the nation as a whole, that those who possess the highest ability—the cleverest of the clever—should be identified as accurately as possible. Of the methods hitherto tried out the so-called 11+ exam has proved to be by far the most trustworthy (1959, p. 117).

Burt's only complaint (1959, p. 32) was that the test and subsequent selection came too late in a child's life.

The system of examination at 11+ and subsequent separation of schools arose in conjunction with a series of official reports issued by government committees during twenty years (the Hadow reports of 1926 and 1931, the Spens report of 1938, the Norwood report of 1943, and the Board of Education's White Paper on Educational Reconstruction—all leading to the Butler Education Act of 1944, which set policy until the mid-1960s when the Labour party vowed to end selection at 11 plus). In the flak surrounding the initial revelation of Burt's fraudulent work, he was often identified as the architect of the 11+ examination. This is not accurate; Burt was not even a member of the various reporting committees, though he did consult frequently with them and he did write extensively for their reports.* Yet it hardly matters whether or not Burt's hand actually moved the pen. The reports embody a particular view of education, clearly identified with the British school of factor analysis, and evidently linked most closely with Cyril Burt's version.

The 11+ examination was an embodiment of Spearmen's hierarchical theory of intelligence, with its innate general factor pervading all cognitive activity. One critic referred to the series of reports as "hymns of praise to the '*g*' factor" (in Hearnshaw, 1979, p. 112). The first Hadow report defined intellectual capacity measured by tests in Burt's favored terms as i.g.c. (innate, general, cognitive) ability: "During childhood, intellectual development progresses as if it were governed largely by a single, central factor, usually known as 'general intelligence,' which may be broadly defined as *innate, all round, intellectual* [my italics for i.g.c.] ability, and appears to enter into everything the child attempts to think, say, or do: this seems the most important factor in determining his work in the classroom."

The 11+ owed its general rationale to the British factorists; in addition, several of its details can also be traced to Burt's school. Why, for example, testing and separation at age eleven? There were practical and historical reasons to be sure; eleven was about the traditional age for transition between primary and secondary schools. But the factorists supplied two important theoretical sup-

*Hearnshaw (1979) reports that Burt had greatest influence over the 1938 Spens report, which recommended sorting at 11 plus and explicitly rejected comprehensive schooling under a single roof thereafter. Burt was piqued at the Norwood report because it downgraded psychological evidence; but, as Hearnshaw notes, this annoyance "masked a basic agreement with the recommendations, which in principle did not differ so much from those of the Spens committee, which he had earlier approved."

ports. First, studies on the growth of children showed that *g* varied widely in early life and first stabilized at about age eleven. Spearman wrote in 1927 (p. 367): "If once, then, a child of 11 years or so has had his relative amount of *g* measured in a really accurate manner, the hope of teachers and parents that he will ever rise to a much higher standing as a late-bloomer would seem to be illusory." Second, Burt's "group factors," which (for purposes of separation by general mental worth) could only be viewed as disturbers of *g*, did not strongly affect a child until after age eleven. The 1931 Hadow report proclaimed that "special abilities rarely reveal themselves in any notable degree before the age of 11."

Burt often claimed that his primary goal in supporting 11+ was a "liberal" one—to provide access to higher education for disadvantaged children whose innate talents might otherwise not be recognized. I do not doubt that a few children of high ability were thus aided, though Burt himself did not believe that many people of high intelligence lay hidden in the lower classes. (He also believed that their numbers were rapidly decreasing as intelligent people moved up the social ladder leaving the lower classes more and more depleted of intellectual talent—1946, p. 15. R. Herrnstein [1971] caused quite a ruckus with the identical argument, recycled, a few years back.)

Yet the major effect of 11+, in terms of human lives and hopes, surely lay with its primary numerical result—80 percent branded as unfit for higher education by reason of low innate intellectual ability. Two incidents come to mind, memories of two years spent in Britain during the regime of 11+: children, already labeled sufficiently by the location of their school, daily walking through the streets of Leeds in their academic uniforms, readily identified by all as the ones who hadn't qualified; a friend who had failed 11+ but reached the university anyway because she had learned Latin on her own, when her secondary modern school did not teach it and universities still required it for entrance into certain courses (how many other working-class teenagers would have had the means or motivation, whatever their talents and desires?).

Burt was committed to his eugenic vision of saving Britain by finding and educating its few people of eminent talent. For the rest, I assume that he wished them well and hoped to match their education with their ability as he perceived it. But the 80 percent were not included in his plan for the preservation of British greatness. Of them, he wrote (1959, p. 123):

> It should be an essential part of the child's education to teach him how to face a possible beating on the 11+ (or any other examination), just as he should learn to take a beating in a half-mile race, or in a bout with boxing gloves, or a football match with a rival school.

Could Burt feel the pain of hopes dashed by biological proclamation if he was willing seriously to compare a permanent brand of intellectual inferiority with the loss of a single footrace?

L. L. Thurstone and the vectors of mind

Thurstone's critique and reconstruction

L. L. Thurstone was born (1887) and bred in Chicago (Ph.D., University of Chicago, 1917, professor of psychology at his alma mater from 1924 to his death in 1955). Perhaps it is not surprising that a man who wrote his major work from the heart of America during the Great Depression should have been the exterminating angel of Spearman's *g*. One could easily construct a moral fable in the heroic mold: Thurstone, free from the blinding dogmas of class bias, sees through the error of reification and hereditarian assumptions to unmask *g* as logically fallacious, scientifically worthless, and morally ambiguous. But our complex world grants validity to few such tales, and this one is as false and empty as most in its genre. Thurstone did undo *g* for some of the reasons cited above, but not because he acknowledged the deeper conceptual errors that had engendered it. In fact, Thurstone disliked *g* because he felt that it was not real enough!

Thurstone did not doubt that factor analysis should seek, as its primary objective, to identify real aspects of mind that could be linked to definite causes. Cyril Burt named his major book *The Factors of the Mind,* Thurstone, who invented the geometrical depiction of tests and factors as vectors (Figs. 6.6, 6.7), called his major work (1935) *The Vectors of Mind.* "The object of factor analysis," Thurstone wrote (1935, p. 53), "is to discover the mental faculties."

Thurstone argued that Spearman and Burt's method of principal components had failed to identify true vectors of mind because it placed factor axes in the wrong geometrical positions.

He objected strenuously both to the first principal component (which produced Spearman's *g*) and to the subsequent components (which identified "group factors" in clusters of positive and negative projections of tests).

The first principal component, Spearman's *g*, is a grand average of all tests in matrices of positive correlation coefficients, where all vectors must point in the same general direction (Fig. 6.4). What psychological meaning can such an axis have, Thurstone asked, if its position depends upon the tests included, and shifts drastically from one battery of tests to another?

Consider Fig. 6.10 taken from Thurstone's expansion (1947) of the *Vectors of Mind.* The curved lines form a spherical triangle on the surface of a sphere. Each vector radiates from the center of the sphere (not shown) and intersects the sphere's surface at a point represented by one of the twelve small circles. Thurstone assumes that the twelve vectors represent tests for three "real" faculties of mind, A, B, and C (call them verbal, numerical, and spatial, if you will). The left set of twelve tests includes eight that primarily measure spatial ability and fall near C; two tests measure verbal ability and lie near A, while two reflect numerical skill. But there is nothing sacrosanct about either the number or distribution of tests in a battery. Such decisions are arbitrary; in fact, a tester usually can't impose a decision at all because he doesn't know, in advance, which tests measure what underlying faculty. Another battery of tests (right side of Fig. 6.10) may happen to include eight for verbal skills and only two each for numerical and spatial ability.

The three faculties, Thurstone believes, are real and invariant in position no matter how many tests measure them in any battery. But look what happens to Spearman's *g*. It is simply the average of all tests, and its position—the x in Fig. 6.10—shifts markedly for the arbitrary reason that one battery includes more spatial tests (forcing *g* near spatial pole C) and the other more verbal tests (moving *g* near verbal pole A). What possible psychological meaning can *g* have if it is only an average, buffeted about by changes in the number of tests for different abilities? Thurstone wrote of *g* (1940, p. 208):

> Such a factor can always be found routinely for any set of positively correlated tests, and it means nothing more or less than the average of all the abilities called for by the battery as a whole. Consequently, it varies

from one battery to another and has no fundamental psychological significance beyond the arbitrary collection of tests that anyone happens to put together. . . . We cannot be interested in a general factor which is only the average of any random collection of tests.

Burt had identified group factors by looking for clusters of positive and negative projections on the second and subsequent principal components. Thurstone objected strenuously to this method, not on mathematical grounds, but because he felt that tests could not have negative projections upon real "things." If a factor represented a true vector of *mind,* then an individual test might either measure that entity in part, and have a positive projection upon the factor, or it might not measure it at all, and have a zero projection. But a test could not have a negative projection upon a real vector of mind:

A negative entry . . . would have to be interpreted to mean that the possession of an ability has a detrimental effect on the test performance. One can readily understand how the possession of a certain ability can aid

6•10 Thurstone's illustration of how the position of the first principal component (the x in both figures) is affected by the types of tests included in a battery.

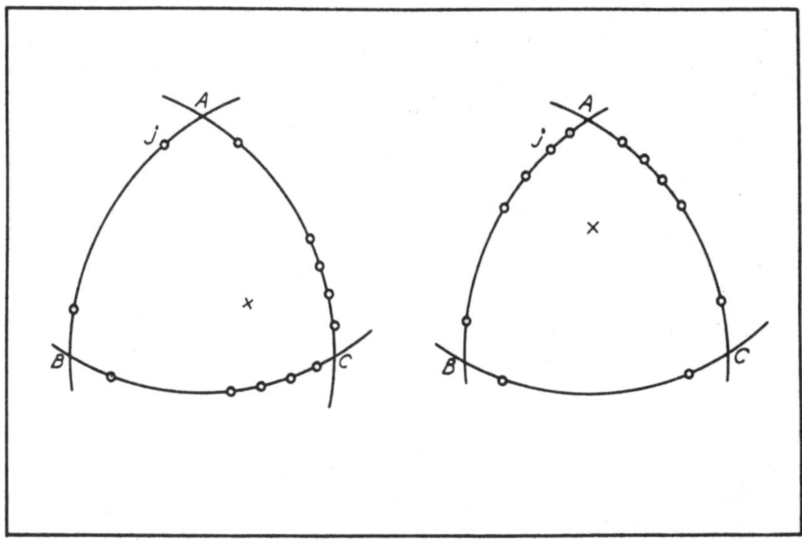

in a test performance, and one can imagine that an ability has no effect on a test performance, but it is difficult to think of abilities that are as often detrimental as helpful in the test performances. Surely, the correct factor matrix for cognitive tests does not have many negative entries, and preferably it should have none at all (1940, pp. 193–194).

Thurstone therefore set out to find the "correct factor matrix" by eliminating negative projections of tests upon axes and making all projections either positive or zero. The principal component axes of Spearman and Burt could not accomplish this because they, perforce, contained all positive projections on the first axis (*g*) and combinations of negative and positive groups on the subsequent "bipolars."

Thurstone's solution was ingenious and represents the most strikingly original, yet simple, idea in the history of factor analysis. Instead of making the first axis a grand average of all vectors and letting the others encompass a steadily decreasing amount of remaining information in the vectors, why not try to place all axes near clusters of vectors. The clusters may reflect real "vectors of mind," imperfectly measured by several tests. A factor axis placed near such a cluster will have high positive projections for tests measuring that primary ability* and zero projections for all tests measuring other primary abilities—as long as the primary abilities are independent and uncorrelated. (Two independent factors are separated by 90° and have zero projection on each other—representing their correlation coefficient of 0.0.)

But how, mathematically, can factor axes be placed near clusters? Here, Thurstone had his great insight. The principal component axes of Burt and Spearman (Fig. 6.6) do not lie in the only position that factor axes can assume. They represent one possible solution, dictated by Spearman's a priori conviction that a single general intelligence exists. They are, in other words, theory-bound, not mathematically necessary—and the theory may be wrong. Thurstone decided to keep one feature of the Spearman-Burt scheme: his factor axes would remain mutually perpendicular, and therefore mathematically uncorrelated. The real vectors of mind, Thurstone reasoned, must represent *independent* primary abilities.

*Thurstone reified his factors, calling them "primary abilities," or "vectors of mind." All these terms represent the same mathematical object in Thurstone's system—factor axes placed near clusters of test vectors.

Thurstone therefore calculated the Spearman-Burt principal components and then *rotated* them to different positions until they lay as close as they could (while still remaining perpendicular) to actual clusters of vectors. In this rotated position, each factor axis would receive high positive projections for the few vectors clustered near it, and zero or near zero projections for all other vectors. When each vector has a high projection on one factor axis and zero or near zero projections on all others, Thurstone referred to the result as a *simple structure.* He redefined the factor problem as a search for simple structure by rotating factor axes from their principal components orientation to positions maximally close to clusters of vectors.

Figs. 6.6 and 6.7 show this process geometrically. The vectors are arranged in two clusters representing verbal and mathematical tests. In Fig. 6.6 the first principal component (*g*) is an average of all vectors, while the second is a bipolar, with verbal tests projecting negatively and arithmetic tests positively. But the verbal and arithmetic clusters are not well defined on this bipolar factor because most of their information has already been projected upon *g,* and little remains for distinction on the second axis. But if the axes are rotated to Thurstone's simple structure (Fig. 6.7), then both clusters are well defined because each is near a factor axis. The arithmetic tests project high on the first simple structure axis and low on the second; the verbal tests project high on the second and low on the first.

The factor problem is not solved pictorially, but by calculation. Thurstone used several mathematical criteria for discovering simple structure. One, still in common use, is called "varimax," or the search for *maximum variance* upon each rotated factor axis. The "variance" of an axis is measured by the spread of test projections upon it. Variance is low on the first principal component because all tests have about the same positive projection, and the spread is limited. But variance is high on rotated axes placed near clusters, because such axes have a few very high projections and other zero or near zero projections, thus maximizing the spread.*

The principal component and simple structure solutions are

*Readers who have done factor analysis for a course on statistics or methodology in the biological or social sciences (quite common in these computer days) will remember something about rotating axes to varimax positions. Like me, they were probably taught this procedure as if it were a mathematical deduction based on the mathematically equivalent; neither is "better." Information is neither gained nor lost by rotating axes; it is merely redistributed. Preferences depend upon the meaning assigned to factor axes. The first principal component demonstrably exists. For Spearman, it is to be cherished as a measure of innate general intelligence. For Thurstone, it is a meaningless average of an arbitrary battery of tests, devoid of psychological significance, and calculated only as an intermediary step in rotation to simple structure.

Not all sets of vectors have a definable "simple structure." A random array without clusters cannot be fit by a set of factors, each with a few high projections and a larger number of near zero projections. The discovery of a simple structure implies that vectors are grouped into clusters, and that clusters are relatively independent of each other. Thurstone continually found simple structure among vectors of mental tests and therefore proclaimed that the tests measure a small number of independent "primary mental abilities," or vectors of mind—a return, in a sense, to an older "faculty psychology" that viewed the mind as a congeries of independent abilities.

> Now it happens, over and over again, that when a factor matrix is found with a very large number of zero entries, the negative entries disappear at the same time. It does not seem as if all this could happen by chance. The reason is probably to be found in the underlying distinct mental processes that are involved in the different tasks. . . . These are what I have called primary mental abilities (1940, p. 194).

Thurstone believed that he had discovered real mental entities with fixed geometric positions. The primary mental abilities (or PMA's as he called them) do not shift their position or change their number in different batteries of tests. The verbal PMA exists in its designated spot whether it is measured by just three tests in one battery, or by twenty-five different tests in another.

> The factorial methods have for their object to isolate the primary abilities by objective experimental procedures so that it may be a question of fact how many abilities are represented in a set of tasks (1938, p. 1).

---

inadequacy of principal components in finding clusters. In fact, it arose historically with reference to a definite theory of intelligence (Thurstone's belief in independent primary mental abilities) and in opposition to another (general intelligence and hierarchy of lesser factors) buttressed by principal components.

Thurstone reified his simple structure axes as primary mental abilities and sought to specify their number. His opinion shifted as he found new PMA's or condensed others, but his basic model included seven PMA's—V for verbal comprehension, W for word fluency, N for number (computational), S for spatial visualization, M for associative memory, P for perceptual speed, and R for reasoning.*

But what had happened to *g*—Spearman's ineluctable, innate, general intelligence—amidst all this rotation of axes? It had simply disappeared. It had been rotated away; it was not there anymore (Fig. 6.7). Thurstone studied the same data used by Spearman and Burt to discover *g*. But now, instead of a hierarchy with a dominant, innate, general intelligence and some subsidiary, trainable group factors, the same data had yielded a set of independent and equally important PMA's, with no hierarchy and no dominant general factor. What psychological meaning could *g* claim if it represented but one possible rendering of information subject to radically different, but mathematically equivalent, interpretations? Thurstone wrote of his most famous empirical study (1938, p. vii):

> So far in our work we have not found the general factor of Spearman. . . . As far as we can determine at present, the tests that have been supposed to be saturated with the general common factor divide their variance among primary factors that are not present in all the tests. We cannot report any general common factor in the battery of 56 tests that have been analyzed in the present study.

The egalitarian interpretation of PMA's

Group factors for specialized abilities have had an interesting odyssey in the history of factor analysis. In Spearman's system they were called "disturbers" of the tetrad equation, and were often purposely eliminated by tossing out all but one test in a cluster—a remarkable way of rendering a hypothesis impervious to disproof. In a famous study, done specifically to discover whether or not

*Thurstone, like Burt, submitted many other sets of data to factor analysis. Burt, chained to his hierarchical model, always found a dominant general factor and subsidiary bipolars, whether he studied anatomical, parapsychological, or aesthetic data. Thurstone, wedded to his model, always discovered independent primary factors. In 1950, for example, he submitted tests of temperament to factor analysis and found primary factors, again seven in number. He named them activity, impulsiveness, emotional stability, sociability, athletic interest, ascendance, and reflectiveness.

group factors existed, Brown and Stephenson (1933) gave twenty-two cognitive tests to three hundred ten-year-old boys. They calculated some disturbingly high tetrads and dropped two tests "because 20 is a sufficiently large number for our present purpose." They then eliminated another for the large tetrads that it generated, excusing themselves by stating: "at worst it is no sin to omit one test from a battery of so many." More high values prompted the further excision of all tetrads including the correlation between two of the nineteen remaining tests, since "the mean of all tetrads involving this correlation is more than 5 times the probable error." Finally, with about one-fourth of the tetrads gone, the remaining eleven thousand formed a distribution close enough to normal. Spearman's "theory of two factors," they proclaimed, "satisfactorily passes the test of experience." "There is in the proof the foundation and development of a scientific experimental psychology; and, although we would be modest, to that extent it constitutes a 'Copernican revolution' " (Brown and Stephenson, 1933, p. 353).

For Cyril Burt, the group factors, although real and important in vocational guidance, were subsidiary to a dominant and innate *g*.

For Thurstone, the old group factors became primary mental abilities. They were the irreducible mental entities; *g* was a delusion.

Copernicus's heliocentric theory can be viewed as a purely mathematical hypothesis, offering a simpler representation for the same astronomical data that Ptolemy had explained by putting the earth at the center of things. Indeed, Copernicus's cautious and practical supporters, including the author of the preface to *De Revolutionibus,* urged just such a pragmatic course in a world populated with inquisitions and indices of forbidden books. But Copernicus's theory eventually produced a furor when its supporters, led by Galileo, insisted upon viewing it as a statement about the real organization of the heavens, not merely as a simpler numerical representation of planetary motion.

So it was with the Spearman-Burt vs. the Thurstone school of factor analysis. Their mathematical representations were equivalent and equally worthy of support. The debate reached a fury of intensity because the two mathematical schools advanced radically different views about the real nature of intelligence—and the acceptance of one or the other entailed a set of fundamental consequences for the practice of education.

With Spearman's *g*, each child can be ranked on a single scale of innate intelligence; all else is subsidiary. General ability can be measured early in life and children can be sorted according to their intellectual promise (as in the 11+ examination).

With Thurstone's PMA's, there is no general ability to measure. Some children are good at some things, others excel in different and independent qualities of mind. Moreover, once the hegemony of *g* was broken, PMA's could bloom like the flowers in spring. Thurstone recognized only a few, but other influential schemes advocated 120 (Guilford, 1956) or perhaps more (Guilford, 1959, p. 477). (Guilford's 120 factors are not induced empirically, but predicted from a theoretical model—represented as a cube of dimensions $6 \times 5 \times 4 = 120$—designating factors for empirical studies to find).

Unilinear ranking of pupils has no place, even in Thurstone's world of just a few PMA's. The essence of each child becomes his individuality, Thurstone wrote (1935, p. 53):

> Even if each individual can be described in terms of a limited number of independent reference abilities, it is still possible for every person to be different from every other person in the world. Each person might be described in terms of his standard scores in a limited number of independent abilities. The number of permutations of these scores would probably be sufficient to guarantee the retention of individualities.

From the midst of an economic depression that reduced many of its intellectual elite to poverty, an America with egalitarian ideals (however rarely practiced) challenged Britain's traditional equation of social class with innate worth. Spearman's *g* had been rotated away, and general mental worth evaporated with it.

One could read the debate between Burt and Thurstone as a mathematical argument about the location of factor axes. This would be as myopic as interpreting the struggle between Galileo and the Church as an argument between two mathematically equivalent schemes for describing planetary motion. Burt certainly understood this larger context when he defended the 11+ examination against Thurstone's assault:

> In educational practice the rash assumption that the general factor has at length been demolished has done much to sanction the impracticable idea that, in classifying children according to their varying capabilities, we need no longer consider their degree of general ability, and have only to allot them to schools of different types according to their special aptitudes; in short, that the examination at 11 plus can best be run on the principle of the caucus-race in Wonderland, where everybody wins and each get some kind of prize (1955, p. 165).

Thurstone, for his part, lobbied hard, producing arguments (and alternate tests) to support his belief that children should not be judged by a single number. He wished, instead, to assess each person as an individual with strengths and weaknesses according to his scores on an array of PMA's (as evidence of his success in altering the practice of testing in the United States, see Guilford, 1959, and Tuddenham, 1962, p. 515).

> Instead of attempting to describe each individual's mental endowment by a single index such as a mental age or an intelligence quotient, it is preferable to describe him in terms of a profile of all the primary factors which are known to be significant. . . . If anyone insists on having a single index such as an I.Q., it can be obtained by taking an average of all the known abilities. But such an index tends so to blur the description of each man that his mental assets and limitations are buried in the single index (1946, p. 110).

Two pages later, Thurstone explicitly links his abstract theory of intelligence with preferred social views.

> This work is consistent not only with the scientific object of identifying the distinguishable mental functions but it seems to be consistent also with the desire to differentiate our treatment of people by recognizing every person in terms of the mental and physical assets which make him unique as an individual (1946, p. 112).

Thurstone produced his fundamental reconstruction without attacking either of the deeper assumptions that had motivated Spearman and Burt—reification and hereditarianism. He worked within established traditions of argument in factor analysis, and reconstructed results and their meaning without altering the premises.

Thurstone never doubted that his PMA's were entities with identifiable causes (see his early work of 1924, pp. 146–147, for the seeds of commitment to reifying abstract concepts—gregariousness in this case—as things within us). He even suspected that his mathematical methods would identify attributes of mind before biology attained the tools to verify them: "It is quite likely that the primary mental abilities will be fairly well isolated by the factorial methods before they are verified by the methods of neurology or genetics. Eventually the results of the several methods of investigating the same phenomena must agree" (1938, p. 2).

The vectors of mind are real, but their causes may be complex and multifarious. Thurstone admitted a strong potential influence for environment, but he emphasized inborn biology:

> Some of the factors may turn out to be defined by endocrinological effects. Others may be defined by biochemical or biophysical parameters of the body fluids or of the central nervous system. Other factors may be defined by neurological or vascular relations in some anatomical locus; still others may involve parameters in the dynamics of the autonomic nervous system; still others may be defined in terms of experience and schooling (1947, p. 57).

Thurstone attacked the environmentalist school, citing evidence from studies of identical twins for the inheritance of PMA's. He also claimed that training would usually enhance innate differences, even while raising the accomplishments of both poorly and well-endowed children:

> Inheritance plays an important part in determining mental performance. It is my own conviction that the arguments of the environmentalists are too much based on sentimentalism. They are often even fanatic on this subject. If the facts support the genetic interpretation, then the accusation of being undemocratic must not be hurled at the biologists. If anyone is undemocratic on this issue, it must be Mother Nature. To the question whether the mental abilities can be trained, the affirmative answer seems to be the only one that makes sense. On the other hand, if two boys who differ markedly in visualizing ability, for example, are given the same amount of training with this type of thinking, I am afraid that they will differ even more at the end of the training than they did at the start (1946, p. 111).

As I have emphasized throughout this book, no simple equation can be made between social preference and biological commitment. We can tell no cardboard tale of hereditarian baddies relegating whole races, classes, and sexes to permanent biological inferiority—or of environmentalist goodies extolling the irreducible worth of all human beings. Other biases must be factored (pardon the vernacular usage) into a complex equation. Hereditarianism becomes an instrument for assigning groups to inferiority only when combined with a belief in ranking and differential worth. Burt united both views in his hereditarian synthesis. Thurstone exceeded Burt in his commitment to a naïve form of reification, and he did not oppose hereditarian claims (though he certainly never pursued them with the single-minded vigor of a Burt). But he chose not to rank and weigh on a single scale of general merit, and his destruction of Burt's primary instrument of ranking—Spearman's *g*—altered the history of mental testing.

Spearman and Burt react

When Thurstone dispersed *g* as an illusion, Spearman was still alive and pugnacious as ever, while Burt was at the height of his powers and influence. Spearman, who had deftly defended *g* for thirty years by incorporating critics within his flexible system, realized that Thurstone could not be so accommodated:

> Hitherto all such attacks on it [*g*] appear to have eventually weakened into mere attempts to explain it more simply. Now, however, there has arisen a very different crisis; in a recent study, nothing has been found to explain; the general factor has just vanished. Moreover, the said study is no ordinary one. Alike for eminence of the author, for judiciousness of plan, and for comprehensiveness of scope, it would be hard to find any match for the very recent work on Primary Mental Abilities by L. L. Thurstone (Spearman, 1939, p. 78).

Spearman admitted that *g*, as an average among tests, could vary in position from battery to battery. But he held that its wandering was minor in scope, and that it always pointed in the same general direction, determined by the pervasive positive correlation between tests. Thurstone had not eliminated *g;* he had merely obscured it by a mathematical dodge, distributing it by bits and pieces among a set of group factors: "The new operation consisted essentially in scattering *g* among such numerous group factors, that the fragment assigned to each separately became too small to be noticeable" (1939, p. 14).

Spearman then turned Thurstone's favorite argument against him. As a convinced reifier, Thurstone believed that PMA's were "out there" in fixed positions within a factorial space. He argued that Spearman and Burt's factors were not "real" because they varied in number and position among different batteries of tests. Spearman retorted that Thurstone's PMA's were also artifacts of chosen tests, not invariant vectors of mind. A PMA could be created simply by constructing a series of redundant tests that would measure the same thing several times, and establish a tight cluster of vectors. Similarly, any PMA could be dispersed by reducing or eliminating the tests that measure it. PMA's are not invariant locations present before anyone ever invented tests to identify them; they are products of the tests themselves:

> We are led to the view that group factors, far from constituting a small number of sharply cut "primary" abilities, are endless in number, indefinitely varying in scope, and even unstable in existence. Any constitutent of ability can become a group factor. Any can cease being so (1939, p. 15).

Spearman had reason to complain. Two years later, for example, Thurstone found a new PMA that he could not interpret (in Thurstone and Thurstone, 1941). He called it X_1 and identified it by strong correlations between three tests that involved the counting of dots. He even admitted that he would have missed X_1 entirely, had his battery included but one test of dotting:

> All these tests have a factor in common; but since the three dot-counting tests are practically isolated from the rest of the battery and without any saturation on the number factor, we have very little to suggest the nature of the factor. It is, no doubt, the sort of function that would ordinarily be lost in the specific variance of the tests if only one of these dot-counting tests had been included in the battery (Thurstone and Thurstone, 1941, pp. 23–24).

Thurstone's attachment to reification blinded him to an obvious alternative. He assumed that X_1 really existed and that he had previously missed it by never including enough tests for its recognition. But suppose that X_1 is a creation of the tests, now "discovered" only because three redundant measures yield a cluster of vectors (and a potential PMA), whereas one different test can only be viewed as an oddball.

There is a general flaw in Thurstone's argument that PMA's are not test-dependent, and that the same factors will appear in any properly constituted battery. Thurstone claimed that an individual test would always record the same PMA's only in simple structures that are "complete and overdetermined" (1947, p. 363)—in other words, only when all the vectors of mind have been properly identified and situated. Indeed, if there *really are* only a few vectors of mind, and if we can know when all have been identified, then any additional test must fall into its proper and unchanging position within the invariant simple structure. But there may be no such thing as an "overdetermined" simple structure, in which all possible factor axes have been discovered. Perhaps the factor axes are not fixed in number, but subject to unlimited increase as new tests are added. Perhaps they are truly test-dependent, and not real underlying entities at all. The very fact that estimates for the number of primary abilities have ranged from Thurstone's 7 or so to Guilford's 120 or more indicates that vectors of mind may be figments of mind.

If Spearman attacked Thurstone by supporting his beloved *g*, then Burt parried by defending a subject equally close to his heart—the identification of group factors by clusters of positive and negative projections on bipolar axes. Thurstone had attacked Spearman and Burt by agreeing that factors must be reified, but disparaging the English method for doing so. He dismissed Spearman's *g* as too variable in position, and rejected Burt's bipolar factors because "negative abilities" cannot exist. Burt replied, quite properly, that Thurstone was too unsubtle a reifier. Factors are not material objects in the head, but principles of classification that order reality. (Burt often argued the contrary position as well—see pp. 288–292.) Classification proceeds by logical dichotomy and antithesis (Burt, 1939). Negative projections do not imply that a person has less than zero of a definite thing. They only record a relative contrast between two abstract qualities of thought. More of something usually goes with less of another—administrative work and scholarly productivity, for example.

As their trump card, both Spearman and Burt argued that Thurstone had not produced a cogent revision of their reality, but only an alternative mathematics for the same data.

> We may, of course, invent methods of factorial research that will always yield a factor-pattern showing some degree of "hierarchical" formation of (if we prefer) what is sometimes called "simple structure." But again the results will mean little or nothing: using the former, we could almost

always demonstrate that a general factor exists; using the latter, we could almost always demonstrate, even with the same set of data, that it does not exist (Burt, 1940, pp. 27–28).

But didn't Burt and Spearman understand that this very defense constituted their own undoing as well as Thurstone's? They were right, undeniably right. Thurstone had not proven an alternate reality. He had begun from different assumptions about the structure of mind and invented a mathematical scheme more in accord with his preferences. But the same criticism applies with equal force to Spearman and Burt. They too had started with an assumption about the nature of intelligence and had devised a mathematical system to buttress it. If the same data can be fit into two such different mathematical schemes, how can we say with assurance that one represents reality and the other a diversionary tinkering? Perhaps both views of reality are wrong, and their mutual failure lies in their common error: a shared belief in the reification of factors.

Copernicus was right, even though acceptable tables of planetary positions can be generated from Ptolemy's system. Burt and Spearman might be right even though Thurstone's mathematics treats the same data with equal facility. To vindicate either view, some legitimate appeal must be made outside the abstract mathematics itself. In this case, some biological grounding must be discovered. If biochemists had ever found Spearman's cerebral energy, if neurologists had ever mapped Thurstone's PMA's to definite areas of the cerebral cortex, then the basis for a preference might have been established. All combatants made appeals to biology and advanced tenuous claims, but no concrete tie has even been confirmed between any neurological object and a factor axis.

We are left only with the mathematics, and therefore cannot validate either system. Both are plagued with the conceptual error of reification. Factor analysis is a fine descriptive tool; I do not think that it will uncover the elusive (and illusory) factors, or vectors, of mind. Thurstone dethroned *g* not by being right with his alternate system, but by being equally wrong—and thus exposing the methodological errors of the entire enterprise.*

*Tuddenham (1962, p. 516) writes: "Test constructors will continue to employ factorial procedures, provided they pay off in improving the efficiency and predictive value of our test batteries, but the hope that factor analysis can supply a short inven-

Oblique axes and second-order g

Since Thurstone pioneered the geometrical representation of tests as vectors, it is surprising that he didn't immediately grasp a technical deficiency in his analysis. If tests are positively correlated, then all vectors must form a set in which no two are separated by an angle of more than 90° (for a right angle implies a correlation coefficient of zero). Thurstone wished to put his simple structure axes as near as possible to clusters within the total set of vectors. Yet he insisted that axes be perpendicular to each other. This criterion guarantees that axes cannot lie really close to clusters of vectors—as Fig. 6.11 indicates. For the maximal separation of vectors is less than 90°, and any two axes, forced to be perpendicular, must therefore lie outside the clusters themselves. Why not abandon this criterion, let the axes themselves be correlated (separated by an angle of less than 90°), and permit them to lie right within the clusters of vectors?

Perpendicular axes have a great conceptual advantage. They are mathematically independent (uncorrelated). If one wishes to identify factor axes as "primary mental abilities," perhaps they had best be uncorrelated—for if factor axes are themselves correlated, then doesn't the cause of that correlation become more "primary" than the factors themselves? But correlated axes also have a different kind of conceptual advantage: they can be placed nearer to clusters of vectors that may represent "mental abilities." You can't have it both ways for sets of vectors drawn from a matrix of positive correlation coefficients: factors may be independent and only close to clusters, or correlated and within clusters. (Neither system is "better"; each has its advantages in certain circumstances. Correlated and uncorrelated axes are both still used, and the argument continues, even in these days of computerized sophistication in factor analysis.)

Thurstone invented rotated axes and simple structure in the early 1930s. In the late 1930s he began to experiment with so-

---

tory of 'basic abilities' is already waning. The continuous difficulties with factor analysis over the last half century suggest that there may be something fundamentally wrong with models which conceptualize intelligence in terms of a finite number of linear dimensions. To the statistician's dictum that whatever exists can be measured, the factorist has added the assumption that whatever can be 'measured' must exist. But the relation may not be reversible, and the assumption may be false."

called oblique simple structures, or systems of correlated axes. (Uncorrelated axes are called "orthogonal" or mutually perpendicular; correlated axes are "oblique" because the angle between them is less than 90°.) Just as several methods may be used for determining orthogonal simple structure, oblique axes can be calculated in a variety of ways, though the object is always to place axes within clusters of vectors. In one relatively simple method, shown in Fig. 6.11, actual vectors occupying extreme positions within the total set are used as factor axes. Note, in contrasting Figs. 6.7 and 6.11, how the factor axes for verbal and mathematical skills have moved from outside the actual clusters (in the orthogonal solution) to the clusters themselves (in the oblique solution).

Most factor-analysts work upon the assumption that correlations may have causes and that factor axes may help us to identify them. If the factor axes are themselves correlated, why not apply

6 • 11 Thurstone's oblique simple structure axes for the same four mental tests depicted in Figs. 6-6 and 6-7. Factor axes are no longer perpendicular to each other. In this example, the factor axes coincide with the peripheral vectors of the cluster.

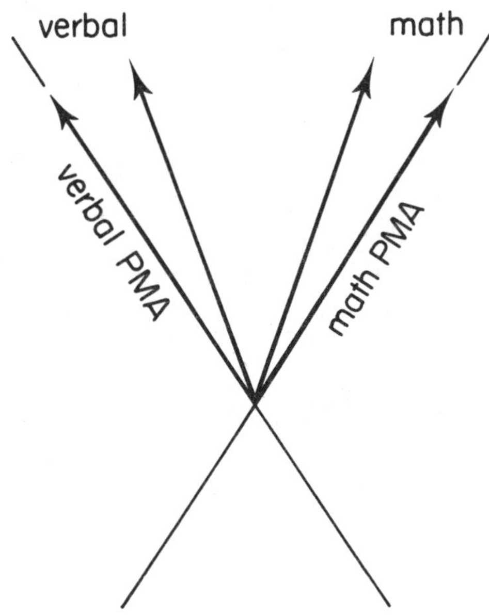

the same argument and ask whether this correlation reflects some higher or more basic cause? The oblique axes of a simple structure for mental tests are usually positively correlated (as in Fig. 6.11). May not the cause of this correlation be identified with Spearman's *g*? Is the old general factor ineluctable after all?

Thurstone wrestled with what he called this "second-order" *g*. I confess that I do not understand why he wrestled so hard, unless the many years of working with orthogonal solutions had set his mind and rendered the concept too unfamiliar to accept at first. If anyone understood the geometrical representation of vectors, it was Thurstone. This representation guarantees that oblique axes will be positively correlated, and that a second-order general factor must therefore exist. Second-order *g* is merely a fancier way of acknowledging what the raw correlation coefficients show—that nearly all correlation coefficients between mental tests are positive.

In any case, Thurstone finally bowed to inevitability and admitted the existence of a second-order general factor. He once even described it in almost Spearmanian terms (1946, p. 110):

> There seems to exist a large number of special abilities that can be identified as primary abilities by the factorial methods, and underlying these special abilities there seems to exist some central energizing factor which promotes the activity of all these special abilities.

It might appear as if all the sound and fury of Thurstone's debate with the British factorists ended in a kind of stately compromise, more favorable to Burt and Spearman, and placing poor Thurstone in the unenviable position of struggling to save face. If the correlation of oblique axes yields a second-order *g*, then weren't Spearman and Burt right all along in their fundamental insistence upon a general factor? Thurstone may have shown that group factors were more important than any British factorist had ever admitted, but hadn't the primacy of *g* reasserted itself?

Arthur Jensen (1979) presents such an interpretation, but it badly misrepresents the history of this debate. Second-order *g* did not unite the disparate schools of Thurstone and the British factorists; it did not even produce a substantial compromise on either side. After all, the quotes I cited from Thurstone on the futility of ranking by IQ and the necessity of constructing profiles based on primary mental abilities for each individual were written after he had admitted the second-order general factor. The two schools were not united and Spearman's *g* was not vindicated for three basic reasons:

1. For Spearman and Burt, *g* cannot merely exist; it must dominate. The *hierarchical* view—with a controlling innate *g* and subsidiary trainable group factors—was fundamental for the British school. How else could unilinear ranking be supported? How else could the 11+ examination be defended? For this examination supposedly measured a controlling mental force that defined a child's general potential and shaped his entire intellectual future.

Thurstone admitted a second-order *g*, but he regarded it as *secondary* in importance to what he continued to call "primary" mental abilities. Quite apart from any psychological speculation, the basic mathematics certainly supports Thurstone's view. Second-order *g* (the correlation of oblique simple structure axes) rarely accounts for more than a small percentage of the total information in a matrix of tests. On the other hand, Spearman's *g* (the first principal component) often encompasses more than half the information. The entire psychological apparatus, and all the practical schemes, of the British school depended upon the preeminence of *g*, not its mere presence. When Thurstone revised *The Vectors of Mind* in 1947, after admitting a second-order general factor, he continued to contrast himself with the British factorists by arguing that his scheme treated group factors as primary and the second-order general factor as residual, while they extolled *g* and considered group factors as secondary.

2. The central reason for claiming that Thurstone's alternate view disproves the necessary reality of Spearman's *g* retains its full force. Thurstone derived his contrasting interpretation from the same data simply by placing factor axes in different locations. One could no longer move directly from the mathematics of factor axes to a psychological meaning.

In the absence of corroborative evidence from biology for one scheme or the other, how can one decide? Ultimately, however much a scientist hates to admit it, the decision becomes a matter of taste, or of prior preference based on personal or cultural biases. Spearman and Burt, as privileged citizens of class-conscious Britain, defended *g* and its linear ranking. Thurstone preferred individual profiles and numerous primary abilities. In an unintentionally amusing aside, Thurstone once mused over the technical differences between Burt and himself, and decided that Burt's propensity for algebraic rather than geometrical representation of factors arose from his deficiency in the spatial PMA:

> The configurational interpretations are evidently distasteful to Burt, for he does not have a single diagram in his text. Perhaps this is indicative of individual differences in imagery types which lead to differences in methods and interpretation among scientists (1947, p. ix).

3. Burt and Spearman based their psychological interpretation of factors on a belief that *g* was dominant *and* real—an innate, general intelligence, marking a person's essential nature. Thurstone's analysis permitted them, at best, a weak second-order *g*. But suppose they had prevailed and established the inevitability of a dominant *g*? Their argument still would have failed for a reason so basic that it passed everybody by. The problem resided in a logical error committed by all the great factorists I have discussed—the desire to reify factors as entities. In a curious way, the entire history that I have traced didn't matter. If Burt and Thurstone had never lived, if an entire profession had been permanently satisfied with Spearman's two-factor theory and had been singing the praises of its dominant *g* for three-quarters of a century since he proposed it, the flaw would be as glaring still.

The fact of pervasive positive correlation between mental tests must be among the most unsurprising major discoveries in the history of science. For positive correlation is the prediction of almost every contradictory theory about its potential cause, including both extreme views: pure hereditarianism (which Spearman and Burt came close to promulgating) and pure environmentalism (which no major thinker has ever been foolish enough to propose). In the first, people do jointly well or poorly on all sorts of tests because they are born either smart or stupid. In the second, they do jointly well or poorly because they either ate, read, learned, and lived in an enriched or a deprived fashion as children. Since both theories predict pervasive positive correlation, the fact of correlation itself can confirm neither. Since *g* is merely one elaborate way of expressing the correlations, its putative existence also says nothing about causes.

Thurstone on the uses of factor analysis

Thurstone sometimes advanced grandiose claims for the explanatory scope of his work. But he also possessed a streak of modesty that one never detects in Burt or Spearman. In reflective moments, he recognized that the choice of factor analysis as a method records the primitive state of knowledge in a field. Factor analysis is a brutally empirical technique, used when a discipline has no firmly established principles, but only a mass of crude data, and a hope that patterns of correlation might provide suggestions for further and more fruitful lines of inquiry. Thurstone wrote (1935, p. xi):

> No one would think of investigating the fundamental laws of classical mechanics by correlational methods or by factor methods, because the laws of classical mechanics are already well known. If nothing were known about the law of falling bodies, it would be sensible to analyze, factorially, a great many attributes of objects that are dropped or thrown from an elevated point. It would then be discovered that one factor is heavily loaded with the time of fall and with the distance fallen but that this factor has a zero loading in the weight of the object. The usefulness of the factor methods will be at the borderline of science.

Nothing had changed when he revised *The Vectors of Mind* (1947, p. 56):

> The exploratory nature of factor analysis is often not understood. Factor analysis has its principal usefulness at the borderline of science. . . . Factor analysis is useful, especially in those domains where basic and fruitful concepts are essentially lacking and where crucial experiments have been difficult to conceive. The new methods have a humble role. They enable us to make only the crudest first map of a new domain.

Note the common phrase—useful "at the borderline of science." According to Thurstone, the decision to use factor analysis as a primary method implies a deep ignorance of principles and causes. That the three greatest factorists in psychology never got beyond these methods—despite all their lip service to neurology, endocrinology, and other potential ways of discovering an innate biology—proves how right Thurstone was. The tragedy of this tale is that the British hereditarians promoted an innatist interpretation of dominant *g* nonetheless, and thereby blunted the hopes of millions.

Epilogue: Arthur Jensen and the resurrection of Spearman's *g*

When I researched this chapter in 1979, I knew that the ghost of Spearman's *g* still haunted modern theories of intelligence. But I thought that its image was veiled, and its influence largely unrecognized. I hoped that a historical analysis of conceptual errors in its formulation and use might expose the hidden fallacies in some contemporary views of intelligence and IQ. I never expected to find a modern defense of IQ from an explicitly Spearmanian perspective.

But then America's best-known hereditarian, Arthur Jensen (1979) revealed himself as an unreconstructed Spearmanian, and centered an eight-hundred-page defense of IQ on the reality of *g*. History often cycles its errors.

Jensen performs most of his factor analyses in Spearman and Burt's preferred principal components orientation (though he is also willing to accept *g* in the form of Thurstone's correlation between oblique simple structure axes). Throughout the book, he names and reifies factors by the usual invalid appeal to mathematical pattern alone. We have *g*'s for general intelligence as well as *g*'s for general athletic ability (with subsidiary group factors for hand and arm strength, hand-eye coordination, and body balance).

Jensen explicitly defines intelligence as "the *g* factor of an indefinitely large and varied battery of mental tests" (p. 249). "We identify intelligence with *g*," he states. "To the extent that a test orders individuals on *g*, it can be said to be a test of intelligence" (p. 224). IQ is our most effective test of intelligence because it projects so strongly upon the first principal component (*g*) in factor analyses of mental tests. Jensen reports (p. 219) that Full Scale IQ of the Wechsler adult scale correlates about 0.9 with *g*, while the 1937 Stanford-Binet projects about 0.8 upon a *g* that remains "highly stable over successive age levels" (while the few small group factors are not always present and tend to be unstable in any case).

Jensen proclaims the "ubiquity" of *g*, extending its scope into realms that might even have embarrassed Spearman himself. Jensen would not only rank people; he believes that all God's creatures can be ordered on a *g* scale from amoebae at the bottom (p. 175) to extraterrestrial intelligences at the top (p. 248). I have not encountered such an explicit chain of being since last I read Kant's speculations about higher beings on Jupiter that bridge the gap between man and God.

Jensen has combined two of the oldest cultural prejudices of Western thought: the ladder of progress as a model for organizing life, and the reification of some abstract quality as a criterion for ranking. Jensen chooses "intelligence" and actually claims that the performance of invertebrates, fishes, and turtles on simple behavioral tests represents, in diminished form, the same essence that humans possess in greater abundance—namely *g*, reified as a measurable object. Evolution then becomes a march up the ladder to realms of more and more *g*.

As a paleontologist, I am astounded. Evolution forms a copiously branching bush, not a unilinear progressive sequence. Jensen speaks of "different levels of the phyletic scale—that is, earthworms, crabs, fishes, turtles, pigeons, rats, and monkeys." Doesn't he realize that modern earthworms and crabs are products of lineages that have probably evolved separately from vertebrates for more than a billion years? They are not our ancestors; they are not even "lower" or less complicated than humans in any meaningful sense. They represent good solutions for their own way of life; they must not be judged by the hubristic notion that one peculiar primate forms a standard for all of life. As for vertebrates, "the turtle" is not, as Jensen claims, "phylogenetically higher than the fish." Turtles evolved much earlier than most modern fishes, and they exist as hundreds of species, while modern bony fishes include almost twenty thousand distinct kinds. What then is "*the* fish" and "*the* turtle"? Does Jensen really think that pigeon-rat-monkey-human represents an evolutionary sequence among warm-blooded vertebrates?

Jensen's caricature of evolution exposes his preference for unilinear ranking by implied worth. With such a perspective, *g* becomes almost irresistible, and Jensen uses it as a universal criterion of rank:

> The common features of experimental tests developed by comparative psychologists that most clearly distinguish, say, chickens from dogs, dogs from monkeys, and monkeys from chimpanzees suggests that they are roughly scalable along a *g* dimension . . . *g* can be viewed as an interspecies concept with a broad biological base culminating in the primates (p. 251).

Not satisfied with awarding *g* a real status as guardian of earthly ranks, Jensen would extend it throughout the universe, arguing that all conceivable intelligence must be measured by it:

> The ubiquity of the concept of intelligence is clearly seen in discussions of the most culturally different beings one could well imagine—extraterrestrial life in the universe. . . . Can one easily imagine "intelligent" beings for whom there is no *g*, or whose *g* is qualitatively rather than quantitatively different from *g* as we know it (p. 248).

Jensen discusses Thurstone's work, but dismisses it as a criticism because Thurstone eventually admitted a second-order *g*. But Jensen has not recognized that if *g* is only a numerically weak, second-order effect, then it cannot support a claim that intelligence is a unitary, dominant entity of mental functioning. I think that Jensen senses his difficulty, because on one chart (p. 220) he calculates both classical *g* as a first principal component and then rotates all the factors (including *g*) to obtain a set of simple structure axes. Thus, he records the same thing twice for each test—*g* as a first principal component and the same information dispersed among simple structure axes—giving some tests a total information of more than 100 percent. Since big *g*'s appear in the same chart with large loadings on simple-structure axes, one might be falsely led to infer that *g* remains large even in simple-structure solutions.

Jensen is contemptuous of Thurstone's orthogonal simple structure, dismissing it as "flatly wrong" (p. 675) and as "scientifically an egregious error" (p. 258). Since he acknowledges that simple structure is mathematically equivalent to principal components, why the uncompromising rejection? It is wrong, Jensen argues, "not mathematically, but psychologically and scientifically" (p. 675) because "it artificially hides or submerges the large general factor" (p. 258) by rotating it away. Jensen has fallen into a vicious circle. He assumes a priori that *g* exists and that simple structure is wrong because it disperses *g*. But Thurstone developed the concept of simple structure largely to claim that *g* is a mathematical artifact. Thurstone wished to disperse *g* and succeeded; it is no disproof of his position to reiterate that he did so.

Jensen also uses *g* more specifically to buttress his claim that the average difference in IQ between whites and blacks records an innate deficiency of intelligence among blacks. He cites the quotation on p. 271 as "Spearman's interesting hypothesis" that blacks score most poorly with respect to whites on tests strongly correlated with *g:*

> This hypothesis is important to the study of test bias, because, if true, it means that the white-black difference in test scores is not mainly attributable to idiosyncratic cultural peculiarities in this or that test, but to a general factor that all the ability tests measure in common. A mean difference between populations that is related to one or more small group factors would seem to be explained more easily in terms of cultural differences than if the mean group difference is most closely related to a broad general factor common to a wide variety of tests (p. 535).

Here we see a reincarnation of the oldest argument in the Spearmanian tradition—the contrast between an innate dominant *g* and trainable group factors. But *g,* as I have shown, is neither clearly a thing, nor necessarily innate if a thing. Even if data existed to confirm Spearman's "interesting hypothesis," the results could not support Jensen's notion of ineluctable, innate difference.

I am grateful to Jensen for one thing: he has demonstrated by example that a reified Spearman's *g* is still the only promising justification for hereditarian theories of mean differences in IQ among human groups. The conceptual errors of reification have plagued *g* from the start, and Thurstone's critique remains as valid today as it was in the 1930s. Spearman's *g* is not an ineluctable entity; it represents one mathematical solution among many equivalent alternatives. The chimerical nature of *g* is the rotten core of Jensen's edifice, and of the entire hereditarian school.

A final thought

> The tendency has always been strong to believe that whatever received a name must be an entity or being, having an independent existence of its own. And if no real entity answering to the name could be found, men did not for that reason suppose that none existed, but imagined that it was something peculiarly abstruse and mysterious.
>
> JOHN STUART MILL

SEVEN

A Positive Conclusion

Walt Whitman, that great man of little brain (see p. 92), advised us to "make much of negatives," and this book has heeded his words, some might say with a vengeance. While most of us can appreciate a cleansing broom, such an object rarely elicits much affection; it certainly produces no integration. But I do not regard this book as a negative exercise in debunking, offering nothing in return once the errors of biological determinism are exposed as social prejudice. I believe that we have much to learn about ourselves from the undeniable fact that we are evolved animals. This understanding cannot permeate through entrenched habits of thought that lead us to reify and rank—habits that arise within social contexts and support them in return. My message, as I hope to convey it at least, is strongly positive for three major reasons.

Debunking as positive science

The popular impression that disproof represents a negative side of science arises from a common, but erroneous, view of history. The idea of unilinear progress not only lies behind the racial rankings that I have criticized as social prejudice throughout this book; it also suggests a false concept of how science develops. In this view, any science begins in the nothingness of ignorance and moves toward truth by gathering more and more information, constructing theories as facts accumulate. In such a world, debunking would be primarily negative, for it would only shuck some rotten apples from the barrel of accumulating knowledge. But the barrel of theory is always full; sciences work with elaborated contexts for explaining facts from the very outset. Creationist biology was dead wrong about the origin of species, but Cuvier's brand of creationism was not an emptier or less-developed world view than Darwin's. Science advances primarily by replacement, not by addition. If the barrel is always full, then the rotten applies must be discarded before better ones can be added.

Scientists do not debunk only to cleanse and purge. They refute older ideas *in the light of* a different view about the nature of things.

Learning by debunking

If it is to have any enduring value, sound debunking must do more than replace one social prejudice with another. It must use more adequate biology to drive out fallacious ideas. (Social prejudices themselves may be refractory, but particular biological supports for them can be dislodged.)

We have rejected many specific theories of biological determinism because our knowledge about human biology, evolution, and genetics has increased. For example, Morton's egregious errors could not be repeated in so bald a way by modern scientists constrained to follow canons of statistical procedure. The antidote to Goddard's claim that a single gene causes feeble-mindedness was not primarily a shift in social preferences, but an important advance in genetical theory—the idea of polygenic inheritance. Absurd as it seems today, the early Mendelians did try to attribute even the most subtle and complex traits (of apolitical anatomy as well as character) to the action of single genes. Polygenic inheritance affirms the participation of many genes—and a host of environmental and interactive effects—in such characters as human skin color.

More importantly, and as a plea for the necessity of biological knowledge, the remarkable lack of genetic differentiation among human groups—a major biological basis for debunking determinism—is a contingent fact of evolutionary history, not an a priori or necessary truth. The world might have been ordered differently. Suppose, for example, that one or several species of our ancestral genus *Australopithecus* had survived—a perfectly reasonable scenario in theory, since new species arise by splitting off from old ones (with ancestors usually surviving, at least for a time), not by the wholesale transformation of ancestors to descendants. We—that is,

Homo sapiens—would then have faced all the moral dilemmas involved in treating a human species of distinctly inferior mental capacity. What would we have done with them—slavery? extirpation? coexistence? menial labor? reservations? zoos?

Similarly, our own species, *Homo sapiens*, might have included a set of subspecies (races) with meaningfully different genetic capacities. If our species were millions of years old (many are), and if its races had been geographically separated for most of this time without significant genetic interchange, then large genetic differences might have slowly accumulated between groups. But *Homo sapiens* is tens of thousands, or at most a few hundred thousand, years old, and all modern human races probably split from a common ancestral stock only tens of thousands of years ago. A few outstanding traits of external appearance lead to our subjective judgment of important differences. But biologists have recently affirmed—as long suspected—that the overall genetic differences among human races are astonishingly small. Although frequencies for different states of a gene differ among races, we have found no "race genes"—that is, states fixed in certain races and absent from all others. Lewontin (1972) studied variation in seventeen genes coding for differences in blood and found that only 6.3 percent of the variation can be attributed to racial membership. Fully 85.4 percent of the variation occurred within local populations (the remaining 8.3 percent records differences among local populations within a race). As Lewontin remarked (personal communication): if the holocaust comes and a small tribe deep in the New Guinea forests are the only survivors, almost all the genetic variation now expressed among the innumerable groups of our four billion people will be preserved.

This information about limited genetic differences among human groups is useful as well as interesting, often in the deepest sense—for saving lives. When American eugenicists attributed diseases of poverty to the inferior genetic construction of poor people, they could propose no systematic remedy other than sterilization. When Joseph Goldberger proved that pellagra was not a genetic disorder, but a result of vitamin deficiency among the poor, he could cure it.

Biology and human nature

If people are so similar genetically, and if previous claims for a direct biological mapping of human affairs have recorded cultural prejudice and not nature, then does biology come up empty as a guide in our search to know ourselves? Are we after all, at birth, the *tabula rasa,* or blank slate, imagined by some eighteenth-century empiricist philosophers? As an evolutionary biologist, I cannot adopt such a nihilistic position without denying the fundamental insight of my profession. The evolutionary unity of humans with all other organisms is the cardinal message of Darwin's revolution for nature's most arrogant species.

We are inextricably part of nature, but human uniqueness is not negated thereby. "Nothing but" an animal is as fallacious a statement as "created in God's own image." It is not mere hubris to argue that *Homo sapiens* is special in some sense—for each species is unique in its own way; shall we judge among the dance of the bees, the song of the humpback whale, and human intelligence?

The impact of human uniqueness upon the world has been enormous because it has established a new kind of evolution to support the transmission across generations of learned knowledge and behavior. Human uniqueness resides primarily in our brains. It is expressed in the culture built upon our intelligence and the power it gives us to manipulate the world. Human societies change by cultural evolution, not as a result of biological alteration. We have no evidence for biological change in brain size or structure since *Homo sapiens* appeared in the fossil record some fifty thousand years ago. (Broca was right in stating that the cranial capacity of Cro Magnon skulls was equal if not superior to ours.) All that we have done since then—the greatest transformation in the shortest time that our planet has experienced since its crust solidified nearly four billion years ago—is the product of cultural evolution. Biological (Darwinian) evolution continues in our species, but its rate, compared with cultural evolution, is so incomparably slow that its impact upon the history of *Homo sapiens* has been small. While the gene for sickle-cell anemia declines in frequency among black Americans, we have invented the railroad, the automobile, radio and television, the atom bomb, the computer, the airplane and spaceship.

Cultural evolution can proceed so quickly because it operates, as biological evolution does not, in the "Lamarckian" mode—by the inheritance of acquired characters. Whatever one generation learns, it can pass to the next by writing, instruction, inculcation, ritual, tradition, and a host of methods that humans have developed to assure continuity in culture. Darwinian evolution, on the other hand, is an indirect process: genetic variation must first be available to construct an advantageous feature, and natural selection must then preserve it. Since genetic variation arises at random, not preferentially directed toward advantageous features, the Darwinian process works slowly. Cultural evolution is not only rapid; it is also readily reversible because its products are not coded in our genes.

The classical arguments of biological determinism fail because the features they invoke to make distinctions among groups are usually the products of cultural evolution. Determinists did seek evidence in anatomical traits built by biological, not cultural, evolution. But, in so doing, they tried to use anatomy for making inferences about capacities and behaviors that they linked to anatomy and we regard as engendered by culture. Cranial capacity per se held as little interest for Morton and Broca as variation in third-toe length; they cared only about the mental characteristics supposedly associated with differences in average brain size among groups. We now believe that different attitudes and styles of thought among human groups are usually the nongenetic products of cultural evolution. In short, the *biological* basis of human uniqueness leads us to reject biological determinism. Our large brain is the biological foundation of intelligence; intelligence is the ground of culture; and cultural transmission builds a new mode of evolution more effective than Darwinian processes in its limited realm—the "inheritance" and modification of learned behavior. As philosopher Stephen Toulmin stated (1977, p. 4): "Culture has the power to impose itself on nature from within."

Yet, if human biology engenders culture, it is also true that culture, once developed, evolved with little or no reference to genetic *variation* among human groups. Does biology, then, play no other valid role in the analysis of human behavior? Is it only a foundation without any insight to offer beyond the unenlightening recognition that complex culture requires a certain level of intelligence?

Most biologists would follow my argument in denying a genetic basis for most behavioral *differences* between groups and for *change* in the complexity of human societies through the recent history of our species. But what about the supposed constancies of personality and behavior, the traits of mind that humans share in all cultures? What, in short, about a general "human nature"? Some biologists would grant Darwinian processes a substantial role not only in establishing long ago, but also in actively maintaining now, a set of specific adaptive behaviors forming a biologically conditioned "human nature." I believe that this old tradition of argument—which has found its most recent expression as "human sociobiology"—is invalid not because biology is irrelevant and human behavior only reflects a disembodied culture, but because human *biology* suggests a different and less constraining role for genetics in the analysis of human nature.

Sociobiology begins with a modern reading of what natural selection is all about—differential reproductive success of individuals. According to the Darwinian imperative, individuals are selected to maximize the contribution of their own genes to future generations, and that is all. (Darwinism is not a theory of progress, increasing complexity, or evolved harmony for the good of species or ecosystems.) Paradoxically (as it seems to many), altruism as well as selfishness can be selected under this criterion—acts of kindness may benefit individuals either because they establish bonds of reciprocal obligation, or because they aid kin who carry copies of the altruist's genes.

Human sociobiologists then survey our behaviors with this criterion in mind. When they identify a behavior that seems to be adaptive in helping an individual's genes along, they develop a story for its origin by natural selection operating upon genetic variation influencing the specific act itself. (These stories are rarely backed by any evidence beyond the inference of adaptation.) Human sociobiology is a theory for the origin and maintenance of *specific, adaptive behaviors* by *natural selection**; these behaviors must

*The brouhaha over sociobiology during the past few years was engendered by this hard version of the argument—genetic proposals (based on an inference of adaptation) for specific human behaviors. Other evolutionists call themselves "sociobiologists," but reject this style of guesswork about specifics. If a sociobiologist is anyone who believes that biological evolution is not irrelevant to human behavior, then I suppose that everybody (creationists excluded) is a sociobiologist. At this point, how-

therefore have a *genetic basis,* since natural selection cannot operate in the absence of genetic variation. Sociobiologists have tried, for example, to identify an adaptive and genetic foundation for aggression, spite, xenophobia, conformity, homosexuality,* and perhaps upward mobility as well (Wilson, 1975).

I believe that modern biology provides a model standing between the despairing claim that biology has nothing to teach us about human behavior and the deterministic theory that specific items of behavior are genetically programed by the action of natural selection. I see two major areas for biological insight:

1. Fruitful analogies. Much of human behavior is surely adaptive; if it weren't, we wouldn't be around any more. But adaptation, in humans, is neither an adequate, nor even a good argument for genetic influence. For in humans, as I argued above (p. 324), adaptation may arise by the alternate route of nongenetic, cultural evolution. Since cultural evolution is so much more rapid than Darwinian evolution, its influence should prevail in the behavioral diversity displayed by human groups. But even when an adaptive behavior is nongenetic, biological analogy may be useful in interpreting its meaning. Adaptive constraints are often strong, and some functions may have to proceed in a certain way whether their underlying impetus be learning or genetic programing.

For example, ecologists have developed a powerful quantitative

---

ever, the term loses its meaning and might as well be dropped. Human sociobiology entered the literature (professional and popular) as a definite theory about the adaptive and genetic basis of specific traits of human behavior. If it has failed in this goal—as I believe it has—then the study of valid relationships between biology and human behavior should receive another name. In a world awash in jargon, I don't see why "behavioral biology" can't extend its umbrella sufficiently to encompass this legitimate material.

*Lest homosexuality seem an unlikely candidate for adaptation since exclusive homosexuals have no children, I report the following story, advocated by E. O. Wilson (1975, 1978). Ancestral human society was organized as a large number of competing family units. Some units were exclusively heterosexual; the gene pool of other units included factors for homosexuality. Homosexuals functioned as helpers to raise the offspring of their heterosexual kin. This behavior aided their genes since the large number of kin they helped to raise held more copies of their genes than their own offspring (had they been heterosexual) might have carried. Groups with homosexual helpers raised more offspring, since they could more than balance, by extra care and higher rates of survival, the potential loss by nonfecundity of their homosexual members. Thus, groups with homosexual members ultimately prevailed over exclusively heterosexual groups, and genes for homosexuality have survived.

theory, called optimal foraging strategy, for studying patterns of exploitation in nature (herbivores by carnivores, plants by herbivores). Cornell University anthropologist Bruce Winterhalder has shown that a community of Cree-speaking peoples in northern Ontario follow some predictions of the theory in their hunting and trapping behavior. Although Winterhalder used a biological theory to understand some aspects of human hunting, he does not believe that the people he studied were genetically selected to hunt as ecological theory predicts they should. He writes (personal communication, July 1978):

> It should go without saying . . . that the causes of human variability of hunting and gathering behavior lie in the socio-cultural realm. For that reason, the models that I used were adapted, not adopted, and then applied to a very circumscribed realm of analysis. . . . For instance, the models assist in analyzing what species a hunter will seek from those available *once a decision has been made to go hunting* [his italics]. They are, however, useless for analyzing why the Cree still hunt (they don't need to), how they decide on a particular day whether to hunt or join a construction crew, the meaning of hunting to a Cree, or any of a plethora of important questions.

In this area, sociobiologists have often fallen into one of the most common errors of reasoning: discovering an analogy and inferring a genetic similarity (literally, in this case!). Analogies are useful but limited; they may reflect common constraints, but not common causes.

2. Biological potentiality vs. biological determinism. Humans are animals, and everything we do is constrained, in some sense, by our biology. Some constraints are so integral to our being that we rarely even recognize them, for we never imagine that life might proceed in another way. Consider our narrow range of average adult size and the consequences of living in the gravitational world of large organisms, not the world of surface forces inhabited by insects (Went, 1968; Gould, 1977). Or the fact that we are born helpless (many animals are not), that we mature slowly, that we must sleep for a large part of the day, that we do not photosynthesize, that we can digest both meat and plants, that we age and die. These are all results of our genetic construction, and all are important influences upon human nature and society.

These biological boundaries are so evident that they have never engendered controversy. The contentious subjects are specific behaviors that distress us and that we struggle with difficulty to change (or enjoy and fear to abandon): aggression, xenophobia, male dominance, for example. Sociobiologists are not genetic determinists in the old eugenical sense of postulating single genes for such complex behaviors. All biologists know that there is no gene "for" aggression, any more than for your lower-left wisdom tooth. We all recognize that genetic influence can be spread diffusely among many genes and that genes set limits to ranges; they do not provide blueprints for exact replicas. In one sense, the debate between sociobiologists and their critics is an argument about the breadth of ranges. For sociobiologists, ranges are narrow enough to program a specific behavior as the predictable result of possessing certain genes. Critics argue that the ranges permitted by these genetic factors are wide enough to include all behaviors that sociobiologists atomize into distinct traits coded by separate genes.

But in another sense, my dispute with human sociobiology is not just a quantitative debate about the extent of ranges. It will not be settled amicably at some golden midpoint, with critics admitting more constraint, sociobiologists more slop. Advocates of narrow and broad ranges do not simply occupy different positions on a smooth continuum; they hold two qualitatively different theories about the biological nature of human behavior. If ranges are narrow, then genes do code for specific traits and natural selection can create and maintain individual items of behavior separately. If ranges are characteristically broad, then selection may set some deeply recessed generating rules; but specific behaviors are epiphenomena of the rules, not objects of Darwinian attention in their own right.

I believe that human sociobiologists have made a fundamental mistake in categories. They are seeking the genetic basis of human behavior at the wrong level. They are searching among the specific products of generating rules—Joe's homosexuality, Martha's fear of strangers—while the rules themselves are the genetic deep structures of human behavior. For example, E. O. Wilson (1978, p. 99) writes: "Are human beings innately aggressive? This is a favorite question of college seminars and cocktail party conversations, and one that raises emotion in political ideologues of all stripes. The answer to it is yes." As evidence, Wilson cites the prevalence of warfare in history and then discounts any current disinclination to fight: "The most peaceable tribes of today were often the ravagers of yesteryear and will probably again produce soldiers and murderers in the future." But if some peoples are peaceable now, then aggression itself cannot be coded in our genes, only the potential for it. If innate only means possible, or even likely in certain environments, then everything we do is innate and the word has no meaning. Aggression is one expression of a generating rule that anticipates peacefulness in other common environments. The range of specific behaviors engendered by the rule is impressive and a fine testimony to flexibility as the hallmark of human behavior. This flexibility should not be obscured by the linguistic error of branding some common expressions of the rule as "innate" because we can predict their occurrence in certain environments.

Sociobiologists work as if Galileo had really mounted the Leaning Tower (apparently he did not), dropped a set of diverse objects over the side, and sought a separate explanation for each behavior—the plunge of the cannonball as a result of something in the nature of cannonballness; the gentle descent of the feather as intrinsic to featherness. We know, instead, that the wide range of different falling behaviors arises from an interaction between two physical rules—gravity and frictional resistance. This interaction can generate a thousand different styles of descent. If we focus on the objects and seek an explanation for the behavior of each in its own terms, we are lost. The search among specific behaviors for the genetic basis of human nature is an example of *biological determinism.* The quest for underlying generating rules expresses a concept of *biological potentiality.* The question is not biological nature vs. nonbiological nurture. Determinism and potentiality are both *biological* theories—but they seek the genetic basis of human nature at fundamentally different levels.

Pursuing the Galilean analogy, if cannonballs act by cannonballness, feathers by featherness, then we can do little beyond concocting a story for the adaptive significance of each. We would never think of doing the great historical experiment—equalizing the effective environment by placing both in a vacuum and observing an identical behavior in descent. This hypothetical example illustrates the social role of biological determinism. It is fundamentally a theory about limits. It takes current ranges in modern environments as an expression of direct genetic programing, rather than a limited display of much broader potential. If a feather acts by featherness, we cannot change its behavior while it remains a feather. If its behavior is an expression of broad rules tied to specific circumstances, we anticipate a wide range of behaviors in different environments.

Why should human behaviorial ranges be so broad, when anatomical ranges are generally narrower? Is this claim for behavioral flexibility merely a social hope, or is it good biology as well? Two different arguments lead me to conclude that wide behavioral ranges should arise as consequences of the evolution and structural organization of our brain. Consider, first of all, the probable adaptive reasons for evolving such a large brain. Human uniqueness lies in the flexibility of what our brain can do. What is intelligence, if not the ability to face problems in an unprogramed (or, as we often say, creative) manner? If intelligence sets us apart among organisms, then I think it probable that natural selection acted to maximize the flexibility of our behavior. What would be more adaptive for a learning and thinking animal: genes selected for aggression, spite, and xenophobia; or selection for learning rules that can generate aggression in appropriate circumstances and peacefulness in others?

Secondly, we must be wary of granting too much power to natural selection by viewing all basic capacities of our brain as direct adaptations. I do not doubt that natural selection acted in building our oversized brains—and I am equally confident that our brains became large as an adaptation for definite roles (probably a complex set of interacting functions). But these assumptions do not lead to the notion, often uncritically embraced by strict Darwinians, that all major capacities of the brain must arise as direct products of natural selection. Our brains are enormously complex computers. If I install a much simpler computer to keep accounts in a factory, it can also perform many other, more complex tasks unrelated to its appointed role. These additional capacities are ineluctable consequences of structural design, not direct adaptations. Our vastly more complex organic computers were also built for reasons, but possess an almost terrifying array of additional capacities—including, I suspect, most of what makes us human. Our ancestors

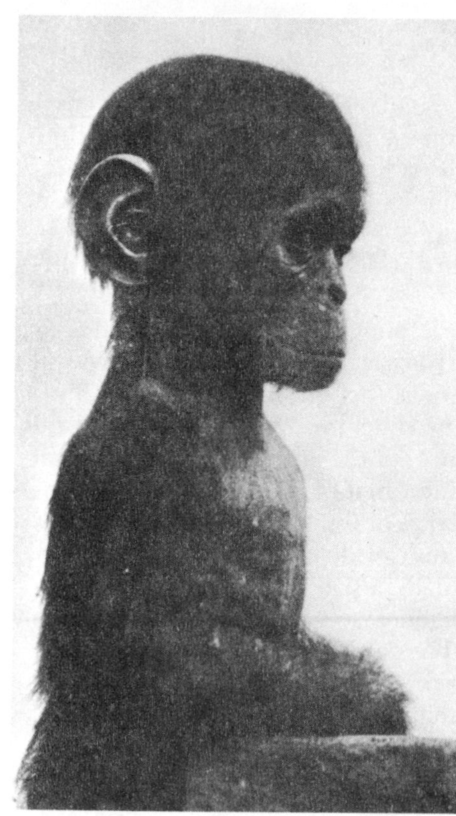

7•1 A juvenile and adult chimpanzee showing the greater resemblance of humans to the baby and illustrating the principle of neoteny in human evolution.

did not read, write, or wonder why most stars do not change their relative positions while five wandering points of light and two larger disks move through a path now called the zodiac. We need not view Bach as a happy spinoff from the value of music in cementing tribal cohesion, or Shakespeare as a fortunate consequence of the role of myth and epic narrative in maintaining hunting bands. Most of the behavioral "traits" that sociobiologists try to explain may never have been subject to direct natural selection at all—and may therefore exhibit a flexibility that features crucial to survival can never display. Should these complex consequences of structural design even be called "traits"? Is this tendency to atomize a behavioral repertory into a set of "things" not another example of the same fallacy of reification that has plagued studies of intelligence throughout our century?

Flexibility is the hallmark of human evolution. If humans evolved, as I believe, by neoteny (see Chapter 4 and Gould, 1977, pp. 352–404), then we are, in a more than metaphorical sense, permanent children. (In neoteny, rates of development slow down and juvenile stages of ancestors become the adult features of descendants.) Many central features of our anatomy link us with fetal and juvenile stages of primates: small face, vaulted cranium and large brain in relation to body size, unrotated big toe, foramen magnum under the skull for correct orientation of the head in upright posture, primary distribution of hair on head, armpits, and pubic areas. If one picture is worth a thousand words, consider Fig. 7.1. In other mammals, exploration, play, and flexibility of behavior are qualities of juveniles, only rarely of adults. We retain not only the anatomical stamp of childhood, but its mental flexibility as well. The idea that natural selection should have worked for flexibility in human evolution is not an ad hoc notion born in hope, but an implication of neoteny as a fundamental process in our evolution. Humans are learning animals.

In T. H. White's novel *The Once and Future King,* a badger relates a parable about the origin of animals. God, he recounts, created all animals as embryos and called each before his throne, offering them whatever additions to their anatomy they desired. All opted for specialized adult features—the lion for claws and sharp teeth, the deer for antlers and hoofs. The human embryo stepped forth last and said:

"Please God, I think that you made me in the shape which I now have for reasons best known to Yourselves and that it would be rude to change. If I am to have my choice, I will stay as I am. I will not alter any of the parts which you gave me. . . . I will stay a defenceless embryo all my life, doing my best to make myself a few feeble implements out of the wood, iron, and the other materials which You have seen fit to put before me. . . ." "Well done," exclaimed the Creator in delighted tone. "Here, all you embryos, come here with your beaks and whatnots to look upon Our first Man. He is the only one who has guessed Our riddle. . . . As for you, Man. . . . You will look like an embryo till they bury you, but all the others will be embryos before your might. Eternally undeveloped, you will always remain potential in Our image, able to see some of Our sorrows and to feel some of Our joys. We are partly sorry for you, Man, but partly hopeful. Run along then, and do your best."

Epilogue

IN 1927 OLIVER WENDELL HOLMES, JR., delivered the Supreme Court's decision upholding the Virginia sterilization law in *Buck v. Bell.* Carrie Buck, a young mother with a child of allegedly feeble mind, had scored a mental age of nine on the Stanford-Binet. Carrie Buck's mother, then fifty-two, had tested at mental age seven. Holmes wrote, in one of the most famous and chilling statements of our century:

> We have seen more than once that the public welfare may call upon the best citizens for their lives. It would be strange if it could not call upon those who already sap the strength of the state for these lesser sacrifices. . . . Three generations of imbeciles are enough.

(The line is often miscited as "three generations of idiots. . . ." But Holmes knew the technical jargon of his time, and the Bucks, though not "normal" by the Stanford-Binet, were one grade above idiots.)

Buck v. Bell is a signpost of history, an event linked with the distant past in my mind. The Babe hit his sixty homers in 1927, and legends are all the more wonderful because they seem so distant. I was therefore shocked by an item in the *Washington Post* on 23 February 1980—for few things can be more disconcerting than a juxtaposition of neatly ordered and separated temporal events. "Over 7,500 sterilized in Virginia," the headline read. The law that Holmes upheld had been implemented for forty-eight years, from 1924 to 1972. The operations had been performed in mental-health facilities, primarily upon white men and women considered feeble-minded and antisocial—including "unwed mothers, prostitutes, petty criminals and children with disciplinary problems."

Carrie Buck, now seventy-two, lives near Charlottesville. Neither she nor her sister Doris would be considered mentally deficient by today's standards. Doris Buck was sterilized under the same law in 1928. She later married Matthew Figgins, a plumber. But Doris Buck was never informed. "They told me," she recalled, "that the operation was for an appendix and rupture." So she and Matthew Figgins tried to conceive a child. They consulted physicians at three hospitals throughout her child-bearing years; no one recognized that her Fallopian tubes had been severed. Last year, Doris Buck Figgins finally discovered the cause of her lifelong sadness.

One might invoke an unfeeling calculus and say that Doris Buck's disappointment ranks as nothing compared with millions dead in wars to support the designs of madmen or the conceits of rulers. But can one measure the pain of a single dream unfulfilled, the hope of a defenseless woman snatched by public power in the name of an ideology advanced to purify a race. May Doris Buck's simple and eloquent testimony stand for millions of deaths and disappointments and help us to remember that the Sabbath was made for man, not man for the Sabbath: "I broke down and cried. My husband and me wanted children desperately. We were crazy about them. I never knew what they'd done to me."

Bibliography

Agassiz, E. C. 1895. *Louis Agassiz: his life and correspondence*. Boston: Houghton, Mifflin, 794pp.

Agassiz, L. 1850. The diversity of origin of the human races. *Christian Examiner* 49: 110–145.

Ashley Montagu, M. F. 1945. Intelligence of northern Negroes and southern whites in the First World War. *American Journal of Psychology* 58: 161–188.

———. 1962. Time, morphology and neoteny in the evolution of man. In *Culture and the evolution of man*, ed. M. F. A. Montagu. New York: Oxford University Press, pp. 324–342.

Bean, Robert Bennett. 1906. Some racial peculiarities of the Negro brain. *American Journal of Anatomy* 5: 353–432.

Binet, A. 1898. Historique des recherches sur les rapports de l'intelligence avec la grandeur et la forme de la tête. *L'Année psychologique* 5: 245–298.

———. 1900. Recherches sur la technique de la mensuration de la tête vivante, plus 4 other memoirs on cephalometry. *L'Année psychologique* 7: 314–429.

———. 1909 (1973 ed.). *Les idées modernes sur les enfants* (with a preface by Jean Piaget). Paris: Flammarion, 232pp.

Binet, A.; and Simon, Th. 1911. *A method of measuring the development of the intelligence of young children*. Lincoln, Illinois: Courier Company, 83pp., 1912.

———. 1916. The development of intelligence in children (The Binet-Simon scale) translated from articles in *L'Année psychologique* from 1905, 1908, and 1911 by Elizabeth S. Kite. Baltimore: Williams and Wilkins, 336pp.

Block, N. J., and Dworkin, G. 1976. *The IQ controversy*. New York: Pantheon.

Blumenbach, J. F. 1825. *A manual of the elements of natural history*. London: W. Simpkin and R. Marshall, 415pp.

Boas, F. 1899. The cephalic index. *American Anthropology* 1: 448–461.

———. 1911. Changes in the bodily form of descendants of immigrants. Senate Document 208, 61st Congress, 2nd Session.

Bolk, L. 1926. *Das Problem der Menschwerdung*. Jena: Gustav Fischer, 44pp.

———. 1929. Origin of racial characteristics in man. *American Journal Physical Anthropology* 13: 1–28.

Borgaonkar, D., and Shah, S. 1974. The XYY chromosome, male—or syndrome. *Progress in Medical Genetics* 10: 135–222.

Bordier, A. 1879. Etude anthropologique sur une série de crânes d'assassins. *Revue d'Anthropologie*, 2nd series, vol. 2, pp. 265–300.

Brigham, C. C. 1923. *A study of American intelligence*. Princeton, N.J.: Princeton University Press, 210pp.

———. 1930. Intelligence tests of immigrant groups. *Psychological Review* 37: 158–165.

Brinton, D. G. 1890. *Races and peoples.* New York: N.D.C. Hodges, 313pp.

Broca, P. 1861. Sur le volume et la forme du cerveau suivant les individus et suivant les races. *Bulletin Société d'Anthropologie Paris* 2: 139–207, 301–321, 441–446.

———. 1862a. Sur les proportions relatives du bras, de l'avant bras et de la clavicule chez les nègres et les européens. *Bulletin Société d'Anthropologie Paris,* vol. 3, part 2, 13pp.

———. 1862b. Sur la capacité des crânes parisiens des diverses époques. *Bulletin Société d'Anthropologie Paris* 3: 102–116.

———. 1862c. Sur les projections de la tête et sur un nouveau procédé de céphalométrie. *Bulletin Société d'Anthropologie Paris* 3: 32pp.

———. 1866. Anthropologie. In *Dictionnaire encyclopédique des sciences médicales,* ed. A. Dechambre. Paris: Masson, pp. 276–300.

———. 1868. *Mémoire sur les crânes des Basques.* Paris: Masson, 79pp.

———. 1873a. Sur les crânes de la caverne de l'Homme-Mort (Lozère). *Revue d'Anthropologie* 2: 1–53.

———. 1873b. Sur la mensuration de la capacité du crâne. *Memoire Société Anthropologie,* 2nd series, vol. 1, 92pp.

———. 1876. *Le programme de l'anthropologie.* Paris: Cusset, 22pp.

Brown, W., and Stephenson, W. A. 1933. A test of the theory of two factors. *British Journal of Psychology* 23: 352–370.

Burt, C. 1909. Experimental tests of general intelligence. *British Journal of Psychology* 3: 94–177.

———. 1912. The inheritance of mental characters. *Eugenics Review* 4: 168–200.

———. 1914. The measurement of intelligence by the Binet tests. *Eugenics Review* 6: 36–50, 140–152.

———. 1921. Mental and scholastic tests. London County Council, 432pp.

———. 1937. *The backward child.* New York: D. Appleton, 694pp.

———. 1939. Lines of possible reconcilement. *British Journal of Pscyhology* 30: 84–93.

———. 1940. *The factors of the mind.* London: University of London Press, 509pp.

———. 1943. Ability and income. *British Journal of Educational Psychology* 13: 83–98.

———. 1946. *Intelligence and fertility.* London: Eugenics Society, 43pp.

———. 1949. The structure of the mind. *British Journal of Educational Psychology* 19: 100–111, 176–199.

———. 1955. The evidence for the concept of intelligence. *British Journal of Educational Psychology* 25: 158–177.

———. 1959. Class differences in general intelligence: III. *British Journal of Statistical Psychology* 12: 15–33.

———. 1959. The examination at eleven plus. *British Journal of Educational Studies* 7: 99–117.

———. 1961. Factor analysis and its neurological basis. *British Journal of Statistical Psychology* 14: 53–71.

———. 1962. Francis Galton and his contributions to psychology. *British Journal of Statistical Psychology* 15: 1–49.

———. 1972. The inheritance of general intelligence. *American Psychology* 27: 175–190.

Bury, J. B. 1920. *The idea of progress.* London: Macmillan, 377pp.

Chase, A. 1977. *The legacy of Malthus.* New York: A. Knopf, 686pp.

Chorover, S. L. 1979. *From genesis to genocide.* Cambridge, MA: Massachusetts Institute of Technology Press.

Combe, G., and Coates, B. H. 1840. Review of *Crania Americana. American Journal of Science* 38: 341–375.

Conway, J. (a presumed alias of Cyril Burt). 1959. Class differences in general intelligence: II. *British Journal of Statistical Psychology* 12: 5–14.

Cope, E. D. 1887. *The origin of the fittest.* New York: Macmillan, 467pp.

———. 1890. Two perils of the Indo-European. *The Open Court* 3: 2052–2054 and 2070–2071.

Count, E. W. 1950. *This is race.* New York: Henry Schuman, 747 pp.

Cox, Catherine M. 1926. The early mental traits of three hundred geniuses. Vol. II. of L. M. Terman (ed.) *Genetic studies of genius.* Stanford, CA: Stanford University Presss, 842pp.

Cravens, H. 1978. *The triumph of evolution: American scientists and the heredity-environment controversy, 1900–1941.* Philadelphia: University of Pennsylvania Press, 351pp.

Cuvier, G. 1812. *Recherches sur les ossemens fossiles.* Vol. 1. Paris: Deterville.

Darwin, C. 1871. *The descent of man.* London: John Murray.

Davenport, C. B.. 1928. Crime, heredity and environment. *Journal of Heredity* 19:307–313.

Dorfman, D. D. 1978. The Cyril Burt question: new findings. *Science* 201: 1177–1186.

Down, J. L. H. 1866. *Observations on an ethnic classification of idiots.* London Hospital Reports, pp. 259–262.

Ellis, Havelock. 1894. *Man and woman.* New York: Charles Scribner's Sons, 561pp.

———.1910. *The criminal.* New York: Charles Scribner's Sons, 440pp.

Epstein, H. T. 1978. Growth spurts during brain development: implications for educational policy and practice. In *Education and the brain,* pp. 343–370, eds. J. S. Chall and A. F. Mirsky. 77th Yearbook, National Society for the Study of Education. Chicago: University of Chicago Press.

Eysenck, H. J. 1953. The logical basis of factor analysis. *American Psychologist* 8: 105–114.

———. 1971. *The IQ argument. Race, intelligence and education.* New York: Library Press, 155pp.

Ferri, E. 1897. *Criminal sociology.* New York: D. Appleton and Company, 284pp.

———. 1911. Various short contributions to criminal sociology. Bericht 7. Internationaler Kongress der Kriminalanthropologie, pp. 49–55, 138–139.

Galton, F. 1884. *Hereditary genius.* New York: D. Appleton, 390pp.

———. 1909. *Memories of my life.* London: Methuen.

Goddard, H. H. 1912. *The Kallikak family, a study in the heredity of feeble-mindedness.* New York: Macmillan, 121pp.

———. 1913. The Binet tests in relation to immigration. *Journal of Psycho-Asthenics* 18: 105–107.

———. 1914. *Feeble-mindedness: its causes and consequences.* New York: MacMillan, 599pp.

———. 1917. Mental tests and the immigrant. *Journal of Delinquency* 2: 243–277.

———. 1917. Review of L. M. Terman, *The Measurement of Intelligence. Journal of Delinquency* 2: 30–32.

———. 1919. *Psychology of the normal and subnormal.* New York: Dodd, Mead and Company, 349pp.

———. 1928. Feeblemindedness: a question of definition. *Journal of Psycho-Asthenics* 33: 219–227.

Gossett, T. F. 1965. *Race: the history of an idea in America.* New York: Schocken Books, 510pp.

Gould, S. J. 1977. *Ever since Darwin.* New York: W.W. Norton.

———. 1977. *Ontogeny and phylogeny.* Cambridge, MA: Harvard University Press.

———. 1978. Morton's ranking of races by cranial capacity. *Science* 200: 503–509.

Guilford, J. P. 1956. The structure of intellect. *Psychological Bulletin* 53: 267–293.

———. 1959. Three faces of intellect. *American Psychology* 14: 469–479.

Hall, G. S. 1904. *Adolescence. Its psychology and its relations to physiology, anthropology, sociology, sex, crime, religion, and education.* 2 vols. New York: D. Appleton and Company, 589 and 784pp.

Haller, J. S., Jr. 1971. *Outcasts from evolution: scientific attitudes of racial inferiority, 1859–1900.* Urbana, Ill.: University of Illinois Press, 228pp.

Hearnshaw, L. S. 1979. *Cyril Burt psychologist.* London: Hodder and Stoughton, 370pp.

Herrnstein, R. 1971. IQ. *Atlantic Monthly,* September, pp. 43–64.

Hervé, G. 1881. Du poids de l'encéphale. *Revue d'Anthropologie,* 2nd séries, vol. 4, pp. 681–698.

Humboldt, A. von. 1849. *Cosmos.* London: H. G. Bohn.

Jarvik, L. F.; Klodin, V.; and Matsuyama, S. S. 1973. Human aggression and the extra Y chromosome: fact or fantasy? *American Psychologist* 28: 674–682.

Jensen, A. R. 1969. How much can we boost IQ and scholastic achievement? *Harvard Educational Review* 33: 1–123.

———. 1979. *Bias in mental testing.* New York: Free Press.

Jerison, J. J. 1973. *The evolution of the brain and intelligence.* New York: Academic Press.

Jouvencel, M. de. 1861. Discussion sur le cerveau. *Bulletin Société d'Anthropologie Paris* 2: 464–474.

Kamin, L. J. 1974. *The science and politics of IQ.* Potomac, MD.: Lawrence Erlbaum Associates.

Kevles, D. J. 1968. Testing the army's intelligence: psychologists and the military in World War I. *Journal of American History* 55: 565–581.

Kidd, B. 1898. *The control of the tropics.* New York: MacMillan, 101pp.

LeBon, G. 1879. Recherches anatomiques et mathématiques sur les lois des variations du volume du cerveau et sur leurs relations avec l'intelligence. *Revue d'Anthropologie,* 2nd series, vol. 2, pp. 27–104.

Linnaeus, C. 1758. *Systema naturae.*

Lippmann, Walter. 1922. The Lippmann-Terman debate. In *The IQ controversy,* eds. N. J. Block and G. Dworkin. New York: Pantheon Books, 1976, pp. 4–44.

Lomax, A., and Berkowitz, N. 1972. The evolutionary taxonomy of culture. *Science* 177: 228–239.

Lombroso, C. 1887. *L'homme criminel.* Paris: F. Alcan, 682pp.

———. 1895. Criminal anthropology applied to pedagogy. *Monist* 6: 50–59.

———. 1896. Histoire des progrès de l'Anthropologie et de la Sociologie criminelles pendant les années 1895–1896. Trav. 4th Cong. Int. d'Anthrop. Crim. Geneva, pp. 187–199.

———. 1911. *Crime: its causes and remedies.* Boston: Little, Brown, 471pp.

Lombroso-Ferrero, G. 1911. Applications de la nouvelle école au Nord de l'Amérique. Bericht 7th Internationaler Kongress der Kriminalanthropologie, pp. 130–137.

Lovejoy, A. O. 1936. *The great chain of being.* Cambridge, MA: Harvard University Press.

Ludmerer, K. M. 1972. *Genetics and American society.* Baltimore, MD.: Johns Hopkins University Press.

Mall, F. P. 1909. On several anatomical characters of the human brain, said to vary according to race and sex, with especial reference to the weight of the frontal lobe. *American Journal of Anatomy* 9: 1–32.

Manouvrier, L. 1903. Conclusions générales sur l'anthropologie des sexes et applications sociales. *Revue de l'École d'Anthropologie* 13: 405–423.

Mark, V., and Ervin, F. 1970. *Violence and the brain*. New York: Harper and Row.

McKim, W. D. 1900. *Heredity and human progress*. New York: G. P. Putnam's Sons, 279pp.

Medawar, P. B. 1977. Unnatural science. *New York Review of Books*, 3 February, pp. 13–18.

Meigs, C. D. 1851. *A memoir of Samuel George Morton, M.D.* Philadelphia: T. K. and P. G. Collins, 48pp.

Montessori, M. 1913. *Pedagogical anthropology*. New York: F. A. Stokes Company, 508pp.

Morton, S. G. 1839. *Crania Americana* or, a comparative view of the skulls of various aboriginal nations of North and South America. Philadelphia: John Pennington, 294pp.

———. 1844. Observations on Egyptian ethnography, derived from anatomy, history, and the monuments [separately reprinted subsequently as *Crania Aegyptiaca*, with title above as subtitle]. *Transactions of the American Philosophical Society* 9: 93–159.

———. 1847. Hybridity in animals, considered in reference to the question of the unity of the human species. *American Journal of Science* 3: 39–50, and 203–212.

———. 1849. Observations on the size of the brain in various races and families of man. *Proceedings of the Academy of Natural Sciences Philadelphia* 4: 221–224.

———. 1850. On the value of the word *species* in zoology. *Proceedings of the Academy of Natural Sciences Philadelphia* 5: 81–82.

———. 1851. On the infrequency of mixed offspring between European and Australian races. *Proceedings of the Academy of Natural Sciences Philadelphia* 5: 173–175.

Myrdal, G. 1944. An American dilemma: the Negro problem and modern democracy. New York: Harper and Brothers, 2 vols., 1483pp.

Newby, I. A. 1969. *Challenge to the court. Social scientists and the defense of segregation, 1954–1966.* Baton Rouge: Louisiana State University Press, 381pp.

Nisbet, R. 1980. *History of the idea of progress*. New York: Basic Books, 370pp.

Nott, J. C., and Gliddon, G. R. 1854. *Types of Mankind*. Philadelphia: Lippincott, Grambo and Company.

———. 1868. *Indigenous races of the earth*. Philadelphia: J. B. Lippincott.

Parmelee, M. 1918. *Criminology*. New York: MacMillan, 522pp.

Pearl, R. 1905. Biometrical studies on man. I. Variation and correlation in brain weight. *Biometrika* 4: 13–104.

———. 1906. On the correlation between intelligence and the size of the head. *Journal of Comparative Neurology and Psychology* 16: 189–199.

Pearl, R., and Fuller, W. N. 1905. Variation and correlation in the earthworm. *Biometrika* 4: 213–229.

Popkin, R. H. 1974. The philosophical basis of modern racism. In *Philosophy and the civilizing arts*, eds. C. Walton and J. P. Anton, pp. 126–165.

Provine, W. B. 1973. Geneticists and the biology of race crossing. *Science* 182: 790–796.

Pyeritz, R.; Schreier, H.; Madansky, C.; Miller, L.; and Beckwith, J. 1977. The XYY male: the making of a myth. In *Biology as a social weapon*, pp. 86–100. Minneapolis: Burgess Publishing Co.

Schreider, E. 1966. Brain weight correlations calculated from original results of

Paul Broca. *American Journal of Physical Anthropology* 25: 153–158.

Serres, E. 1860. Principes d'embryogénie, de zoogénie et de teratogénie *Mémoire de l'Académie des Sciences* 25: 1–943.

Sinkler, G. 1972. *The racial attitudes of American presidents from Abraham Lincoln to Theodore Roosevelt.* New York: Doubleday Anchor Books, 500pp.

Spearman, C. 1904. General intelligence objectively determined and measured. *American Journal of Psychology* 15: 201–293.

———. 1914. The heredity of abilities. *Eugenics Review* 6: 219–237.

———. 1914. The measurement of intelligence. *Eugenics Review* 6: 312–313.

———. 1923. *The nature of "intelligence" and the principles of cognition.* London: MacMillan, 358pp.

———. 1927. *The abilities of man.* New York: MacMillan, 415pp.

———. 1931. Our need of some science in place of the word "intelligence." *Journal of Educational Psychology* 22: 401–410.

———. 1937. *Psychology down the ages.* London: MacMillan, 2 vols., 454 and 355pp.

———. 1939. Determination of factors. *British Journal of Psychology* 30: 78–83.

———. 1939. Thurstone's work re-worked. *Journal of Educational Psychology* 30: 1–16.

Spearman, C., and Wynn Jones, Ll. 1950. *Human ability.* London: MacMillan, 198pp.

Spencer, H. 1895. *The principles of sociology.* 3rd ed. New York: D. Appleton and Company.

Spitzka, E. A. 1903. A study of the brain of the late Major J. W. Powell. *American Anthropology* 5: 585–643.

———. 1907. A study of the brains of six eminent scientists and scholars belonging to the American Anthropometric Society, together with a description of the skull of Professor E. D. Cope. *Transactions of the American Philosophical Society* 21: 175–308.

Stanton, W. 1960. *The leopard's spots: scientific attitudes towards race in America 1815–1859.* Chicago: University of Chicago Press, 245pp.

Stocking, G. 1973. *From chronology to ethnology. James Cowles Prichard and British Anthropology 1800–1850.* In facsimile of 1813 ed. of J. C. Prichard, Researches into the physical history of man. Chicago: University of Chicago Press, pp. ix–cxvii.

Strong, J. 1900. Expansion under new world-conditions. New York: Baker and Taylor, 310pp.

Sully, James. 1895. Studies of childhood. XIV. The child as artist. *Popular Science* 48: 385–395.

Taylor, I.; Walton, P.; and Young, J. 1973. *The new criminology: for a social theory of deviance.* London: Routledge and Kegan Paul, 325pp.

Terman, L. M. 1906. Genius and stupidity. A study of some of the intellectual processes of seven "bright" and seven "stupid" boys. *Pedagogical Seminary* 13: 307–373.

———. 1916. *The measurement of intelligence.* Boston: Houghton Mifflin, 362pp.

Terman, L. M., and 12 others. 1917. *The Sanford Revision extension of the Binet-Simon scale for measuring intelligence.* Baltimore: Warwick and York, 179pp.

Terman, L. M. 1919. *The intelligence of school children.* Boston: Houghton Mifflin, 317pp.

Terman, L. M., and 5 others. 1923. *Intelligence tests and school reorganization.* Yonkers-on-Hudson, N.Y.: World Book Company, 111pp.

Terman, L. M., and Merrill, Maud A. 1937. *Measuring intelligence. A guide to the administration of the new revised Stanford-Binet tests of intelligence*. Boston: Houghton Mifflin, 461pp.

Thomson, G. H. 1939. *The factorial analysis of human ability*. Boston: Houghton Mifflin.

Thorndike, E. L. 1940. *Human nature and the social order*. New York: Macmillan, 1019pp.

Thurstone, L. L. 1924. *The nature of intelligence*. London: Kegan Paul, Trench, Trubner and Company, 167pp.

———. 1935. *The vectors of mind*. Chicago: University of Chicago Press, 266pp.

———. 1938. *Primary mental abilities*. Chicago: University of Chicago Press, Psychometric Monographs, no. 1, 121pp.

———. 1940. Current issues in factor analysis. *Psychological Bulletin* 37: 189–236.

———. 1946. Theories of intelligence. *Scientific Monthly*, February, pp. 101–112.

———. 1947. *Multiple factor analysis*. Chicago: University of Chicago Press, 535pp.

———. 1950. The factorial description of temperament. *Science* 111: 454–455.

Thurstone, L. L., and Thurstone, T. G. 1941. *Factorial studies of intelligence*. Chicago: University of Chicago Press, Psychometric Monographs, no. 2, 94pp.

Tobias, P. V. 1970. Brain-size, grey matter, and race—fact or fiction? *American Journal of Physical Anthropology* 32: 3–26.

Todd, T. W., and Lyon, D. W., Jr. 1924. Endocranial suture closure. Its progress and age relationship. Part 1. Adult males of white stock. *American Journal of Physical Anthropology* 7: 325–384.

———. 1925a. Cranial suture closure. II. Ectocranial closure in adult males of white stock. *American Journal of Physical Anthropology* 8: 23–40.

———. 1925b. Cranial suture closure. III. Endocranial closure in adult males of Negro stock. *American Journal of Physical Anthropology* 8: 47–71.

Topinard, P. 1878. *Anthropology*. London: Chapman and Hall, 548pp.

———. 1887. L'anthropologie criminelle. *Revue d'Anthropologie*, 3rd series, vol. 2: 658–691.

———. 1888. Le poids de l'encéphale d'après les registres de Paul Broca. *Mémoires Société d'Anthropologie Paris*, 2nd series, vol. 3, pp. 1–41.

Toulmin, S. 1977. Back to nature. *New York Review of Books*, 9 June, pp. 3–6.

Tuddenham, R. D. 1962. The nature and measurement of intelligence. In *Psychology in the making*, ed. L. Postman, pp. 469–525. New York: Alfred A. Knopf.

Vogt, Carl. 1864. *Lectures on man*. London: Longman, Green, Longman, and Roberts, 475pp.

Voisin, F. 1843. *De l'idiotie chez les enfants*. Paris: J.-B. Ballière.

Washington, B. T. 1904. *Working with the hands*. New York: Doubleday, Page and Company, 246pp.

Went, F. W. 1968. The size of man. *American Scientist* 56: 400–413.

Weston, R. F. 1972. *Racism in U.S. imperialism: the influence of racial assumptions on American foreign policy 1893–1946*. Columbia: University of South Carolina Press, 291pp.

Wilson, E. O. 1975. *Sociobiology*. Cambridge, MA: Harvard University Press.

———. 1978. *On human nature*. Cambridge, MA: Harvard University Press.

Wilson, L. G. 1970. *Sir Charles Lyell's scientific journals on the species question*. New Haven: Yale University Press, 572pp.

Wolfle, Dael. 1940. *Factor analysis to 1940*. Psychometric Monographs No. 3, Psychometric Society. Chicago: University of Chicago Press, 69pp.

Yerkes, R. M. 1917a. The Binet version versus the point scale method of measuring intelligence. *Journal of Applied Psychology* 1: 111–122.

———. 1917b. How may we discover the children who need special care. *Mental Hygiene* 1: 252–259.

Yerkes, R. M. (ed.) 1921. Psychological examining in the United States army. *Memoirs of the National Academy of Sciences,* vol. 15, 890pp.

Yerkes, R. M. 1941. Man power and military effectiveness: the case for human engineering. *Journal of Consulting Psychology* 5: 205–209.

Zimmern, H. 1898. Criminal anthropology in Italy. *Popular Science Monthly* 52: 743–760.

Index